# Sustainable Urban Futures

Series Editor
Zaheer Allam
Le Hochet
Morcellement Raffray
Terre-Rouge, Mauritius

This series includes a broad range of Pivot length books offering accessible and applied texts designed to appeal to both practitioners and academics in the field. Pivots in the series will explore how sustainability can be achieved in Future Cities and how technology can assist in supporting sustainable transitions to better respond to the urgencies of climate change, equity needs and inclusivity aligning the two core themes of Urban Science and Future Science.

More information about this series at
http://www.palgrave.com/gp/series/16631

Zaheer Allam

# The Rise of Autonomous Smart Cities

Technology, Economic Performance and Climate Resilience

palgrave
macmillan

Zaheer Allam
Le Hochet
Morcellement Raffray
Terre-Rouge, Mauritius

ISSN 2730-6607                    ISSN 2730-6615    (electronic)
Sustainable Urban Futures
ISBN 978-3-030-59447-3        ISBN 978-3-030-59448-0    (eBook)
https://doi.org/10.1007/978-3-030-59448-0

Cover pattern © Melisa Hasan

This Palgrave Macmillan imprint is published by the registered company Springer Nature
Switzerland AG.
The registered company address is: Gewerbestrasse 11, 6330 Cham, Switzerland

# Series Editor Preface

Sustainable Urban Futures is an exciting and innovative Pivot series from Palgrave Macmillan providing readers with concise and accessible coverage of how cities will change and develop in response to the emerging issues of the day. The rapid growth of cities is giving rise to a range of economic, social, and environmental problems. The challenge for decisionmakers is how to manage growth sustainably, tapping the potential benefits of urbanisation while avoiding its exclusionary and environmentally damaging tendencies. This series is highly relevant within the current global context, providing guidance for how cities can tackle the simultaneous challenges of climate change, public health crisis, and mass urbanisation. Each book contributes to the discussions on innovative planning, design, and policy solutions leading to urban transformation, and to the extent to which these solutions are effective in achieving immediate and long-term sustainability goals. The authors of this book series, which consist of both academics and practitioners, offer perspectives from the following disciplines: urban planning and design, human geography, economics, behavioural sciences, environmental psychology, and other policy-related disciplines. The geographical scope of the book series is global, with inter- and trans-disciplinary contributions relating to urban interventions from countries within both the Global North and South.

Terre-Rouge, Mauritius                          Zaheer Allam and Sina Shahab

# CONTENTS

# ABOUT THE AUTHOR

**Zaheer Allam** holds a PhD in Humanities, a Master of Arts (Res) in Political Economy, an MBA and a Bachelor of Applied Science in Architectural Science from universities in Australia and the United Kingdom. Based in Mauritius, he works as an Urban Strategist for The Port Louis Development Initiative and consults for the Global Creative Leadership Initiative on a number of projects on the thematic of Smart Cities and on strategies dwelling in the increasing role of technology in Culture and the Society. Allam is also the African Representative of the International Society of Biourbanism (ISB), member of the Advisory Circle of the International Federation of Landscape Architects (IFLA) and a member of a number of other international bodies. For his contributions to society, he was elevated, by the President of Mauritius, to the rank of Officer of the Order of the Star and Key of the Indian Ocean (OSK), the highest distinct order of Merit in Mauritius, and he is a recipient of a number of other awards. He is the author of numerous peer-reviewed publications and author of eight books on the subject of Future Cities.

# CHAPTER 1

# Introducing the Concept of Autonomous City

**Abstract** Cities have gained over the years from the adoption of technology and this has been reflected in the emergence and support of numerous concepts like sustainable, eco, resilient, amongst others, until the most recent one—the smart city. A commonality among those concepts is the understanding that technology can aid in increasing the performance and efficiency of urban areas, on which ICT corporations are tapping into for economic benefits. One avenue, which this book supports, is that the concept will undeniably evolve from the dimension of data gathering to that of action response—based on the data gathered and its interpretation—giving rise to the concept of 'Autonomous Cities'. This introductory chapter explores the concept of how data gathering pursuits will evolve to that of automating action-based computer generated decisions in cities.

**Keywords** Smart city • Future cities • Autonomous city
Sustainability • Resilience • Liveability • Technology

## INTRODUCTION

The digital revolution has brought about numerous changes in the urban realm. In turn, these changes have brought about tangible solutions to a range of issues ranging from social, economic, political and environmental issues. For instance, in the social realm, there is much progress in

© The Author(s) 2021
Z. Allam, *The Rise of Autonomous Smart Cities*, Sustainable Urban
Futures, https://doi.org/10.1007/978-3-030-59448-0_1

1

achieving inclusivity and equity in terms of housing, provision of services like waste management, distribution of resources like water, energy and in provision of health services. Economically, the digital revolution has inspired numerous innovations like car sharing—which is now widely adopted across various geographies and amongst others, it has inspired the emergence of digital financial transaction, influenced new innovations in the tourism industry, in the conservation and consumption of cultural heritage and art to name a few. These revolutions, in turn, have influenced the creation of new businesses, provided new opportunities for job creation, and allowed local and national governments to expand their revenue bases. On the environmental front, there are numerous initiatives like the adoption of alternative energy, the greening of cities, the ability to predict and monitor weather conditions in real-time among other things that are helping conserve the environment.

Those positive impacts have been experienced progressively, even as urban concepts continued to change; moving from the initial concept of Global city, to Sustainable City, the Resilient City, to the Creative City and to, most recently, Smart Cities (Hatuka et al. 2018). The Smart City concept is seen to be the most progressive of those concepts, and it has been around for a few years. For instance, it is argued that it started in 2009 in the city of Santander, Spain with over 20,000 sensors distributed throughout its urban landscape (Hartwig 2020). However, its adoption has been slow, and has been criticized for various reasons. One reason that critiques have cited is the emphasis on making everything new in the city. This ideology goes against the concept of sustainability and the need for connectedness that are attributed to smart cities. That is, in the pursuit of newness, Jacobs (1961) has shown that most well-thought traditional infrastructures, building and streets among others are compromised. In most cases, they are demolished to give way for new developments and this renders cities into 'human traps' as they end up with no physical linkage between people; thus, constricting human physical relationships, which can be further accentuated with the aid of technology. For instance, McSpadden (2018) argues that in a city like Rio De Janeiro, Brazil, a population of over 13 million people is controlled by just a few people sitting in a room full of screens showing images obtained from thousands of cameras and sensors installed a cross the city. On this, Weber (2019) highlights that, of importance, urban managers need to be concentrating on making urban residents smarter, and this would help achieve concept of smart city quicker, that it would when relying on data.

While its adoption has shown potentials in various areas, it remains a prominent challenge for policy makers to implement the Smart City concept. In most cases, this has been left in the hands of ICT corporations whose missions and goals are seen to primarily maximize their revenues. Thereby, they ensure that the smart products they install in most cities are customized to ride on individual platforms that seldom support open data sharing; thus, making it hard for product cohabitation among various systems. This has then led to increased calls for standardization of protocols and networks to allow the unlocking the true potential for smart cities (Allam and Jones 2020; Allam 2019). Doing this would allow for unfettered installation of emerging technologies like IoT and 5G among others. With many of these, Norman (2018) highlights that the emergence of a new concept 'Autonomous Cities' may be in the offing, and with this, cities would not only collect data from the various smart devices and technologies but would also have the ability to process them for insights and in some cases, enact on those decisions. However, this concept will need bridging on already existing barriers both in legislation, policy, and politics as this will mean giving the city the power to remediate on emerging issues and concerns without the need for human interventions.

While automating cities would have numerous benefits, including increasing efficiencies, increasing potential to install smart technologies, and providing new business frontiers like in the tourism sectors, they would have some negatives that need to be addressed. First, as noted above, such a concept needs to be regulated by having ample and appropriate legislations and policies that would among other things define the boundaries within which such a concept can be actualized. This, as noted by Frank et al. (2018), is important as automating cities may lead to the compromise of numerous humanistic aspects like liveability, human physical interactions and the 'wholeness' that should come along with built fractals. On this, (Kamel Boulos et al. 2015) add that automation of cities may in some instances lead to mechanical environments that only increase human seclusion and dilute the intended outcomes of having a vibrant and liveable urban environment. Hawksworth et al. (2018) argue that automation would also have some negative bearing in the employment sector, with a sizeable number of people employed in areas like waste management, transport sectors, and other service industries facing the reality of being rendered jobless. This point affirms the need for proactive legislation and policies that would ensure that as those people—being replaced in their job areas by machines, the urban environment, are provided with a robust mechanism allowing them to transition to other, or secure new, job opportunities.

However, on the positive side, automation of urban fabric is hailed for the potential it holds in areas like addressing transportation challenges that numerous cities face, especially due to the high urban population that are always on the move. The concept has the potential in helping urban areas solve the ever increasing waste management challenges, and in turn help promote areas like circular economy, which in the long-run, would help in better addressing joblessness by encouraging job transitions after the automation of this sector. The environment is also among the main beneficiaries of automation, as the circular economy that will be promoted as a result will help in, among other things, minimizing waste, emphasize the adoption of renewable sources and promote upcycling and recycling of materials. Anttiroiko et al. (2014) note that the achievement of such benefits would be accentuated, especially with the application of Artificial Intelligent (AI) technologies that are at the heart of smart city concepts, and also that are expected to play a key role in automated cities concepts.

One prominent area that automation of city is expected to play a key role is in addressing the challenge of climate change. Murali (2020) notes that through digitization, cities are becoming smart in areas like energy consumption, waste management, resource consumption, compact building among others, and these are expected to get better in automated cities. To sum the benefits, improvements and efficiencies in those areas are expected to have positive bearings in reduction of emissions, destruction of natural resources and excessive exploitation of the ecology. By achieving these, urban areas are then seen as source of the solution to climate change. For instance, in the recent past, it has become normal for people to work from home due to the availability of cutting-edge technologies, allowing for video conferencing, and remote access to workplaces. Such approaches reduce vehicular activities, which in turn means a reduction in the amount of emissions, and energy consumption; all to the good of the environment.

The issue of climate change, however, is intricate and would require concerted efforts from different stakeholders. For cities, the issue of connectivity needs to be robust and void of issues like unstandardized protocols and networks that make it hard for devices to communicate or share data. Therefore this book emphasizes the need for enactment of legislations and policies that would help lift the lid in data sharing. Such also need to ensure that contracted ICT corporations are not after their own profit-oriented interests, but are in tandem with the objectives of helping urban areas achieve the goal of being smart, more so by not only installing said technologies, but also by getting returns, beyond monetary forms, from such investments. In addition, the concept of Smart Cities being pursued should assist cities be more cohesive, inclusive, sustainable, and safe as anticipated in the SDG 11 and in the New Urban Agenda document.

Some of the ideas (like those of automated cities) discussed in this book may seem futurist, but, they may be well close to being actualized; hence, underlining the urgent need for discussions on the matter. On this, this book has taken heed to introduce the topic, discuss the numerous potentials the use of technology has on the actualization of Smart City concept and highlighted some of the potential areas of concerns. In this regard, this book is organized such that the Chap. 2 discusses how different technologies like Big Data, AI and others are aiding in the rise of autonomous Smart Cities. Chapter 3 discusses in depth how those technologies help achieve complexity and connectivity in the Smart cities which Chap. 4 delve deeper in showing how achieving the 'Smart' agenda in a city help in uplifting the economic aspect of the city and its residents and how that could be affected if the concept of capitalism, more so advanced by large ICT corporation is not overcome. Chapter 5 dwells on how autonomous smart cities are helping address the challenge of climate change while Chap. 6 discusses how the concept of Smart City could be exploited as a tool to advance tourism agenda, and in the regeneration and exploitation of cultural tools for the benefits of the residents and the economic welfare of the city.

## REFERENCES

Allam, Z. (2019). The Emergence of Anti-Privacy and Control at the Nexus between the Concepts of Safe City and Smart City. *Smart Cities, 2.* https://doi.org/10.3390/smartcities2010007

Allam, Z., & Jones, D. S. (2020). On the Coronavirus (COVID-19) Outbreak and the Smart City Network: Universal Data Sharing Standards Coupled with Artificial Intelligence (AI) to Benefit Urban Health Monitoring and Management. *Healthcare, 8,* 46.

Anttiroiko, A. V., Valkama, P., & Bailey, S. J. (2014). Smart Cities in the New Service Economy: Building Platforms for Smart Services. *Artificial Intelligence and Society, 29,* 323–334.

Frank, M. R., Sun, L., Cebrian, M., Youn, H., & Rahwan, I. (2018). Small Cities Face Greater Impact from Automation. *Journal of the Royal Society Interface, 15,* 20170946. https://doi.org/10.1098/rsif.2017.0946

Hartwig, M. (2020). Smart Cities. Retrieved July 5, 2020, from https://www. bmw.com/en/innovation/smart-cities.html.

Hatuka, T., Rosen-Zvi, I., Birnhack, M., Toch, E., & Zur, H. (2018). The Political Premises of Contemporary Urban Concepts: The Global City, the Sustainable City, the Resilient City, the Creative City, and the Smart City. *Planning Theory & Practice, 19*, 160–179. https://doi.org/10.1080/14649357.2018.1455216

Hawksworth, J., Berriman, R., & Goel, S. (2018). *Will Robots Really Steal Our Jobs?: An International Analysis of the Potential Long Term Impact of Automation.* UK: PWC.

Jacobs, J. (1961). *The Death and Life of Great American Cities.* New York, NY: Vintage Books.

Kamel Boulos, M. N., Tsouros, A. D., & Holopainen, A. (2015). 'Social, Innovative and Smart Cities are Happy and Resilient': Insights from the WHO EURO 2014 International Healthy Cities Conference. *International Journal of Health Geographics, 14*, 3. https://doi.org/10.1186/1476-072X-14-3

McSpadden, K. (2018). 5 Critiques of the Smart City Push. Retrieved July 10, 2020, from https://e27.co/5-critiques-smart-city-push-20180907/.

Murali, S. (2020). Automation Makes it Possible for Cities to go Green. Retrieved July 10, 2020, from https://www.smartcitiesdive.com/news/automation-makes-it-possible-for-cities-to-go-green/569672/.

Norman, B. (2018). Are Autonomous Cities Our Urban Future? *Nature Communications, 9*, 1–3.

Weber, V. (2019). Smart Cities Must Pay More Attention to the People Who Live in Them. Retrieved July 10, 2020, from https://www.weforum.org/agenda/2019/04/why-smart-cities-should-listen-to-residents/.

# CHAPTER 2

# Big Data, Artificial Intelligence and the Rise of Autonomous Smart Cities

**Abstract** The roles of technology in enhancing the performance and efficiency of cities are increasingly hailed as the Smart Cities' model gains recognition and acceptance in various urban quarters. This is accentuated with the increasing wealth of data made available via Big Data, mined from an array of sensors tapping into complex Internet of Things (IoT) networks. The rise in data, and computational power to interpret it at near real time capabilities, allows urban leaders to better understand the complexities related to the urban realm, and to be better equipped to take urban decisions. However, as urban fabrics are increasingly connected, there is today the possibility of automating certain urban processes in response to urban challenges; but this should be carefully led so as not to allow the creation of fully automated cities that render mechanical decisions made to impact negatively on urban liveability.

**Keywords** Big Data • Artificial Intelligence • Smart Cities • Autonomous cities • Sustainability • Identity

© The Author(s) 2021
Z. Allam, *The Rise of Autonomous Smart Cities*, Sustainable Urban Futures, https://doi.org/10.1007/978-3-030-59448-0_2

7

## Introduction

In the wake of the ongoing fourth industrial revolution, there are spirited efforts to adopt technologically advanced urban concepts with the aim of addressing eminent urban issues. In particular, to address the lingering challenges brought about by the rapid increase in urban population with that of the increasing rates of urbanization. On these, different urban concepts that hinges on the power of technology have been tried in different cities (Yavuz et al. 2018; United Nations 2016; Van Winden and van den Buuse 2017; Kamel Boulos et al. 2015; Oh and Phillips 2014). A majority of those, which include the concepts of technoparks, Eco-cities -which Singapore has extensively exploited and gained in popularity in the last decades, and Smart Cities relies heavily on the power of technology to aptly collect, store, analyze and transmit large amounts of data from different urban quarters, the environment and the city populace. Yigitcanlar and Bulu (2016) explain how different technologies such as Artificial Intelligence (AI)(Payne 2018; Scarcello 2018), Internet of Things (IoT) (Naganathan and Rao 2018; Patel and Patel 2016; Al-Fugaha et al. 2015), Big Data (Barkham et al. 2018; Bhadani 2016; Batty 2016), and Crowd Computing are now adopted by engineers, urban planners and managers to collect valuable data essential for primary urban management. Similarly, they also have the ability to interpret data collected from Internet of Things (IOT) devices installed in cities along with data shared on social media by the urban dwellers (Bassoo et al. 2018; Obinikpo and Kantarci 2017; Abaker et al. 2016; Allam 2020f, g), and thus, manage to act according with both the capacity of efficient and inclusive management.

Ersoy and Alberto (2019) highlight how this wealth in data is allowing urban leaders to take better informed decisions, which in turn have facilitated the adoption of strategies that have resulted in increased efficiency and quality performance of cities. Bibri and Krogstie (2019) share that availability of the vast data from different quarters has also been handy in promoting equitable urban environment and sustainability efforts. With the novel application of the Big Data and AI technologies, Mahdavinejad et al. (2018) highlight that a score of cities have managed to counter security concerns as such technologies have the capacities to allow facial recognition and motion sensing through different sensors and devices employed in cities. Another area that has been positively impacted by the availability of data is the optimal consumption of resources and use alternative sources of such especially energy, which have had significant contribution in climate change (Fan et al. 2019; Zhou et al. 2016; Kylili and Paris 2015). With data gathering and its interpretation, it has been possible to implement alternative and renewable energy policies as explained by Zhou et al. (2016). Similarly, it has been possible to equitably provide essential services like waste management, water supply and sewerage services to households, businesses and industries (Al Nuaimi et al. 2015).

In pursuit of technology application in urban planning and development, the concept of Smart Cities has emerged perhaps as the most recognized and is being embraced across the globe, with different Smart City policies being implemented (Van Winden and van den Buuse 2017; Shepard 2017; Ishida 2017; Danigelis 2017; Slavova and Okwechime 2016; Richter et al. 2015). The popularity of this concept is pegged on numerous factors, but its ability to accommodate diverse technologies is unmatched. In particular, noting that cities have been undergoing numerous and momentous challenges as explained by Taylor Buck and While (2015), the digital solutions available in this concept have been seen to serve as the new 'silver' bullet for numerous urban challenges. Indeed, on this, a report by EY (2014) highlights that the concept augurs well with the urban population which have also turned to be increasingly digital; therefore, a combination of smart technologies and digitally connected people brings the best of this concept. This truth is revealed in the increased demand for unique and high-tech urban products, including urban infrastructures, buildings and different governments and private systems.

In addition, the above is also boosted by the increased interests and efforts by large ICT corporations that have been seen to intensively engage in R&D focusing on Smart Cities' digital solutions, and also their interests to participate in actualizing these concepts. On this, it is noted that large corporations such as IBM Corporation, Cisco, Tesla, Google, Huawei, Schneider Electric, Ericsson, Intel and many others are tirelessly working toward providing high quality 'smart' branded technologies (Allam and Dhunny 2019). Besides these large corporations, a report by McKinsey (McKinsey & Company 2018) highlights that small scaled startups are also actively helping actualize the Smart City concept through diverse digital technologies and through various concepts like biking and bike services, online shopping, taxi services and mobile money services to name a few (Marzano et al. 2019). The availability of these numerous 'smart' solutions, as noted by Huikkola and Kohtamäki (2019), is possible due to the advancement in technology and the commitment by different corporations, despite their size to invest heavily in R&D.

Such activities in R&D have prompted an increase in products geared towards the automation of different urban processes, and these have had positive impacts in increasing efficiency in service delivery. Such automation like online money and mobile transfer, transportation startups like autonomous vehicles, and Smart cameras and smart lighting technologies amongst many others are just a few marvels that Smart Cities have been associated with (Allam 2020b, c). But, as these are being celebrated, there are some highlighted concerns, especially touching on ethics and moral aspects of these technology in rendering cities as mechanical environments and stifling the humanistic aspect of urban fabrics. That is, despite their potential in making urban fabric flawless, such efforts have been seen to distance human interaction and there are worries that people in such cities are already trapped in technology and intricate architectural marvels, thus, increasing their seclusion. On this, Kamel Boulos et al. (2015) highlight that human beings are social and that aspect cannot be replaced by anything else, thus, they need to maintain contact and foster a rich interaction.

In cities, for them to be seen as truly liveable, the increase in digital solutions applications should be rendered such as way that they promote human interactions. Newman (2010) showcases that this is possible as some cities, such as the island state Singapore, have managed to tap into the power of these technologies to secure green spaces that serve as recreation areas where human interactions are espoused. This example showcases how the automation of different urban fabrics, when remotely connected, have the capacity to lead to a responsive network that promote a balanced metabolism of the world. This have been seen to be especially important as the world is currently facing a climate emergency, and the optimal use of resources could potentially allow cities to remain within the emission limits set in different international conventions focusing on the environment and climate change.

## THE SMART CITY MARKET

The advent of Smart Cities has been beneficial for the technology industry as this has created an insatiable market for digital products, devices and services that are continually demanded in different cities around the world (Allam 2018, 2019, 2020a, 2020d, 2020e, 2020h; Allam and Newman 2018; Lim et al. 2018). As explained above, this concept has attracted the attention of large ICT Corporations which are seen to position themselves to respond to this demand with the aim to tap into the economic benefits that are ubiquitous to this industry. The reason behind the urgency by ICT corporations to invest heavily in this industry is captured in a report by 'Zion Market Research' which highlights that the Smart City Market globally accounted over $ 955.3 billion in 2017, and at an estimated average growth rate of 16%, the figure would grow to over 2,700.1 billion by 2024. Another report by Grand View Research (2019) showcased that the Smart Cities market has been growing in reap and bounds, from $71.3 billion as of 2018 to clock over $237.6 billion by 2025, at a compounded growth rate of 18.9%. Further still, Singh (2019) believes that the market value of this industry has been higher at $308.0 billion as of 2018 and is expected to grow over $717.2 by 2023. These three reports and the diversity in the figures captured serve as a pointer at how lucrative this industry is, and the reason why numerous ICT corporations; both established ones and start-ups are in tireless motion to engage in this market (Richter et al. 2015).

The attractiveness of the mouthwatering growth in the market size of this industry is that the competition therein is health, and has unprecedented potential to benefit the local municipalities and ultimately the population of these cities, especially in terms of costs and quality of digital products that are availed to them. This is more so as most of the corporations have been seen to invest substantially in the R&D for the development and deployment of products in cities, in what has been posited as branding of ICT goods. This branding, has in turn resulted in boosting the Smart City concept, and different Smart Cities around the globe are now identifiable by the uniqueness that have resulted into their branding. Novelli (2005) explains that by this branding, cities have spurred such economic fronts like place branding and tourism in different forms ranging from place tourism, medical tourism, education tourism and cultural tourism to name a few.

A case in point is the Songdo Smart City in South Korea; a new smart city which is taunted as an ubiquitous-Eco City that encompasses smart infrastructures worth of an International Business District (Kolotouchkina and Seisdedos 2017). Another notable example is the case of Singapore, Bangkok and Thailand which have been riding on the concept of sustainability warranted by the Smart City concept to brand position themselves. Using this strategy, as showcased by Taecharungroj et al. (2019), these cities have managed to endear the global community, and in return, they have positively overturned their economic fortunes. In particular, they are said to have put more emphasis on environmental sustainability, and it is not surprising to find sizeable green spaces, numerous alternative energy sources being used, clean rivers and water ways and buildings with green walls and roofs.

## THE WEB OF INTERNET OF THINGS

The above branding concepts are made possible following adoption of a wide range of advanced technologies, and in particular, the use of Internet of Things (IoT), which principally revolves around the central idea that devices can be seamlessly connected among one another to help exchange information. Those products today form part of our everyday lives and are used to collect a vast array of data to aid in assessing both lifestyle and health.

The reasoning behind IoT, being a pivotal technological component of Smart Cities, is its ability to allow different infrastructures to be connected, and the data collected from the various devices that ride on this platform either wirelessly or wired can be used to render smart decisions (Guo et al. 2018; Gil et al. 2016; Evans 2011). This means that any device that has the capacity to connect to the internet is a candidate for use in Smart Cities and this is great breakthrough, as the more the number of devices, the more the data collected, and ultimately, the quality of information and decision. On this, from the literature, as showcased by Tzafestas (2018), one challenge that Smart Cities have been facing is the lack of a standardized protocol through which different components and devices can interact, but with IoT, this challenge has been partly surmounted (Zanella et al. 2014; Noura et al. 2019; Minoli et al. 2017).

The popularity of this platform is depicted by the increasing number of devices and components used in different smart cities (Silva et al. 2018). According to Park et al. (2018), there were 1.6 billion such devices, and by 2018, their numbers had increased to over 3.3 billion, which is a growth rate of approximately 43%. Going forward, as more smart cities are being implemented, the numbers of such IoT components and devices is expected to continue growing to over 25 billion by 2025 as projected by Rishi and Saluja (2019) from EY. This will translate to a global market revenue of approximately $1.1 trillion. Such figures will mean that Smart Cities have the potential to be characterized with more complex structures, as advocated by Salingaros (Salingaros 2014, 2000). Such complexity, built from a more interconnected web of IoT devices, means efficiency, real-time response to issues, and improved communication and social interaction, which depict a health urban area, especially in respect to live-ability when viewed using the social lens.

The application of IoT is not without some hurdles and is somehow seen to be under exploited. First, though it noted that any device that has the capacity to connect to the internet qualify to play a part in the Smart City concept, such devices are not always on the same network infrastructure. That is, some are still using the old 3G, while others are on 4G and some have the capability to operate on the 5G network infrastructures. Such lack of standardization means that communication amongst devices is restricted; hence, the expected output may not be achieved (Noura et al. 2019). Taylor Buck and While (2017) explain that partly, this problem is exacerbated by the practices of ICT Corporation wanting to yield competitive market advantages, thus, ensuring that its devices are uniquely different from those of its competitors. Unfortunately, such practices derail the success of Smart Cities and make it somehow expensive.

Despite these issues, it is believed that the 5G network, that is already in application in some cities like Chicago, Detroit, Edinburgh, London, Seoul, Incheon, Shenzhen and Al Rayyan city in Qatar (Fisher 2019; Lomas 2019; Garcia 2019) and being advanced by companies like Verizon, Sprint, Unicom, Vodacom and EE mobile network operators amongst many others and also by Huawei in its mobile devices will allow for increased connectivity. Minoli et al. (2017) express that this network will cater for the proliferation of IoT components and devices; rendering a substantial, and previously unseen, amount of urban data for analysis. This will be possible because the 5G network allows for an increase in data sharing at speed 100 times faster than the popular 4G LTE network (Goldstein 2019), while being ultra-reliable and allowing for denser connections -as opposed to the preceding. It is believed that a shift to this kind of network will actualize the smart city experience into a true real-time sharing of data and information; hence, such sectors like transport and security dockets are expected to be greatly enhanced. Cheng et al. (2018) further highlight that the 5G network will also be instrumental in achieving smart manufacturing; thus, positively impacting on the sustainability efforts in cities, especially noting that this one sector has contributed greatly to climate change. Therefore, as is advanced in the succeeding sections, there is much that Smart Cities are to gain from IoTs; especially when it is shrewdly integrated with other advanced technologies like AI, Big Data and a better and faster network.

## ARTIFICIAL INTELLIGENCE AND BIG DATA

The rise of IoT as discussed above has brought numerous advancements, accelerated with the increase in amount of data collected and analyzed. As Bibri (2018) explains, this has subsequently led to a better overview of the processes of the urban fabric. According to him, IoT is highly associated with Big Data analytics; hence, has been instrumental in the optimization of energy efficiency, improved urban security, efficiency in traffic sector and in the climate change mitigation processes. But as mentioned above, IoT is just one out of the many technologies that are rendering the Smart City concept into a reality, and in all fairness, it cannot be assumed as a standalone technology and need to be used alongside other advanced technologies like Crowd Computing, AI, Big Data and Blockchain to maximum the benefits derived from it. For this reason, especially in need to cope with the rise in data, such technologies like AI and Big Data have allowed for the better processing and analysis of data for real-time response arising from a diverse range of urban issues.

Lozada et al. (2019) highlights how Big Data technologies is essential in helping urban managers gain priceless insights from the massive data that is garnered from the numerous smart IoT components: sensors, cameras and other related devices. On this, Popovič et al. (2018) explain that it would be practically valueless to have IoT devices and components installed without having proper technologies to handle the massive data output from such; hence, making the Big Data technology a critical component in the analysis endeavor. Indeed, the impact of Big Data technology is so important that Aydiner et al. (2019) opines that it has become a leading target area amongst top executives in different fields and also in academia who believe it holds the key to advancing current and future operations of different sectors, and those who leverage its potential are better positioned to reap maximum competitive advantage. This is true even in urban areas that are confronted by increasing challenges ranging from surging energy demand, traffic issues, housing challenges and sustainability problems amongst other things.

Besides Big Data technologies, Artificial Intelligence (AI) is another critical dimension that helps in obtaining even better insights from the analysis of big data. According to IBM, AI has the potential to recreate human capabilities and intelligence into computers, and this quality make it a super component as it allows for informed inferences, perceptions and predictions on technically complex fields like weather forecasting; that have had far reaching impacts on both humans and the environment. One key characteristic of AI that make it a perfect match for the other Smart Cities technologies is its ability to help in the automation of processes. That is, allowing for substitution of human efforts in most of data collection, storage, analysis and transmission of results into different nodes, thus, reducing some obvious biases and errors.

Botta et al. (2016) discuss that AI has the capacity to allow different facets of life to trend in uncharted grounds like the use of robotics in health care and AI algorithms in weather prediction and in other fields. Its potentials have been demonstrated in different popular cultures like in films, movies and books, where it has been branded as one of the marvels of our modern days in how much it can help in different spheres. Nevertheless, as Erler (2019) highlights, AI, like all the other technologies, do not have stand-alone capabilities as depicted in some of the popular cultures. The truth is that it only processes data as per designated commands-in very specifically designed fashions, and vaporizing fears that in in the immediate future, AI can be harmful to mankind. Indeed, Zorins and Grabusts (2015) highlight that AI is still in its infancy stages and it would take substantive effort, time and investment for it to mimic human capabilities, more so since it relies on data sets thrown to it, and it only executes such. Therefore, it needs to be integrated with other technologies for it to achieve the desired smart urban outcomes.

Through the gathering of data from diverse urban sectors, and its collective processing using the technologies discussed above and others like crowd computing and Blockchain, it is possible to detect emerging patterns, which can provide a blueprint to render better and informed decisions in urban management. Da Cruz et al. (2019) express that this is critical since most urban areas are facing some of what can be termed as twenty first century challenges: like climate change, increased urban population, urbanization, security threats and unstable and fragile economies, which can only be addressed by having absolute information. The challenge of financial constraints is another major one that warrants the need for the above technologies to be employed. On this, Richards and Thompson (2019) explain that most urban areas receive little or no financial support from national agencies despite the increase in scope of responsibilities that these continue to attract more people migrating to cities, and as they also confront the consequences of climate change. With informed planning blueprints, Freudendal-Pedersen et al. (2019) express that cities have the potential to overcome such financial constraints by automating some urban processes like energy production, especially through the adoption of alternative energy sources and smart grids.

Automation is also feasible in the manufacturing sector as expressed by Cheng et al. (2018), and here, cities would benefit from optimal use of resources and reduced waste generation. In the transportation and traffic sector, which have been a major challenge in most of the urban areas, automation of some of processes especially by the adoption of smart technologies like AI is would be major breakthrough overcoming some challenges like congestion, emissions, accidents and insecurity as expressed by Taeihagh and Lim (2019). Joss et al. (2019) highlight that such automation would be beneficial to the optimal utilization of available infrastructures, as well as increase efficiency and speed-up service delivery in urban areas. As is showcased in the section below, the availability of AI, Big Data and other technologies like Blockchain holds the key to this automation, and the benefits to be derived thereof are unsurmountable, though, with some hurdles like insecurity and rendering the city as mechanical that need to be overcome.

## AUTOMATING THE CITY

Coupling AI with execution capabilities has allowed various industries to turn towards automation, leading to enhanced manufacturing and operation processes. On this, it cannot be underestimated of how much AI can help achieve in the manufacturing sector. Loucks and Hupfer (2018) acknowledge that AI can be viewed as one of the missing links in different industries, especially in helping them overcome some obvious internal challenges like limited number of expertise, integration, overload of information and critical decision making. They argue that AI has been seen to transform how different activities and processes are executed through various ways like direct automation and continuous operation as the robots powered by AI do not became lethargic like human beings; hence, they can run as long as they are required to do so. Also, AI allows for safer operational environments to be created in industries, as most of the dangerous, risky, laborious and time intensive activities that are handled by machines; which are not susceptible, like humans, to workplace accidents (Jarrahi 2018). It is also understood to contribute greatly in minimizing operation costs following things like reduction in number of workforces, and its capability to run continuously. With such, industries and organizations also benefit from increased scalability with the outcomes characterized with high quality and innovative products—also produced in a predetermined time frame.

In terms of statistical contribution, different reports have showcased the potential of AI, not only in manufacturing industry, but on the global GDP contribution. According to the ITU (2018), the employment of AI in different sectors has the potential to yield an additional global revenue of approximately $13 trillion by 2030. Another report by Accenture (Purdy and Daugherty 2017) highlights that such revenue would be equivalent to approximately $14 trillion by 2035. On this, PWC (2018) have also weighted in and argue that by 2030, the global economy could reach a higher of $104 trillion higher from the current estimated GDP of $88.1 trillion, and with the contribution of the AI, the figure could go higher by 14% which is equivalent to $15.7 trillion. These promising statistics from the contribution of AI are credited to the increased uptake of this technology in both developed and developing economies, and they are real even though AI threaten job security in different industrial sectors across the globe. But, a look at the positive side of it, as described by Purdy and Daugherty (2017) shows that the replacement of humans by automated machines and robots needs consideration, as these are allowing for cost reduction, efficiency and increased productivity among other benefits cited above.

Just like industries are turning towards the revamping of their infrastructure to allow for the collection of data and robots for the execution of automated commands, in the near future, a majority of cities will be seen to have the same underlying infrastructure. That is, IoT for the collection of data, AI for the processing of this data, and execution commands for the output in various sectors. Indeed, in a way, such trends have already been achieved, though in smaller and limited scales in some sectors in several urban areas. For instance, in the transportation sector, there are some road networks that are now controlled by smart technologies seamlessly in real time in control centers; allowing for more fluidity in traffic as seen in Dallas, Pittsburgh, Farmington Hills, Michigan, Sydney and New York amongst many others (Abril 2018; Snow 2017; Australian Government 2018). Citron (2019) highlights how these technologies are allowing for installation of video detection and radar to monitor traffic, and, in real time, adjust signals thus, ensuring no traffic build-up. Besides reducing traffic, employment of these technologies are being hailed for reducing travel time and more importantly, reducing vehicular emissions; hence, aligning with calls from various quarters to sustainably reduce emissions by 2030 (Drell 2011). Buying from the success achieved in cities where such technologies have been installed, a full-scale adoption in a majority of cities across the globe would have significant economic, social and environmental impacts, and the urban revenue would even surpass projected targets.

Besides the traffic sector, AI and IoT have also had significant and unmatched benefits in the energy sector. From vast literature (UN Environment 2019; IRENA 2018; Bergman 2018; Aliyu and Amadu 2017; Ochola 2018; Carreón and Worrell 2018), the energy sector is seen to have the greatest impact in the contribution of emissions, and any effort to alter or reduce this would be very welcome. With the above smart technologies, Dutta Pramanik et al. (2018) express that the future of cities, and that of the environment has some hope on achieving sustainability to ensure the survival of humankind. Ferrández-Pastor et al. (2018) further express that the IoT and AI infrastructure are significant in facilitating integration and monitoring of renewable energies, more so due to the data collection and processing capacity. Automation of this sector would mean optimal use of the energy, and the energy resources and such would have significant impacts in different urban infrastructure (Zhou et al. 2016; Al Nuaimi et al. 2015). Ferrero Bermejo et al. (2019) highlight that the Artificial Neural Networks (ANN), which is a form of AI is especially significant in enhancing real-time estimation of assets and predicting the future energy demand and thus, help in informing the adoption of renewable energy.

Despite the myriad of benefits that these technologies have, their applicability is still in the infancy stages, with numerous challenges and hurdles that will need to be overcome. But with the recent surge in the Smart Cities concept, such technologies have the opportunities to be expansively employed, and this will increase the interest especially on large corporation to invest even more on R&D on these technologies; thus, increasing their ubiquitousness. One such areas that this is taking shape is the advancement in network connectivity where now, most large corporations and other stakeholders are rooting for the 5G network that is almost being accelerated in different regions. According to Condoluci and Mahmoodi (2018), this is seen as a blessing to the reinforcement of AI and IoT technologies since it offers unmatched connectivity speed, flexibility, reconfigurability, programmability and increased bandwidth, which are prerequisite for a functional smart city and the said technologies.

## Automated Cities and the Optimization of Climate Change Mitigation

Unlike some decades back, this epoch is facing many challenges despite an increase in technological application. Among these challenges is the climate change, which is threatening major facets of the globe, and unsurprisingly, it is being experienced with no region, country or economy being immune (Chancel and Piketty 2015; Diffenbaugh and Burke 2019; Carlos et al. 2017). The challenge of climate change has prompted the formulation of numerous mitigation plans and legal instruments focusing on ensuring the survival of both biodiversity and people (Alfredsson et al. 2018; REN21 2018; Tu 2017). Majorly, the survival of these entities is threatened by increasing global temperatures, rising ocean levels, diminishing snow and ice and environmental degradation. The climate has changed and became almost unpredictable, with extremes like strong hurricanes and cyclones, increased precipitation and flooding being a few of the pointers to this. The United Nations (2019) underscores that the above challenges can be directly linked to human contribution, especially in $CO_2$ and other GHG emissions. And, as it is widely shared above, these have their roots mostly on urban areas where consumption of energy intensive products and services is rampant and in high demand (Purkus et al. 2017; Ortiz et al. 2018; Papavasiliou and Oren 2014; Ma et al. 2018). Similarly, it is there that the conversion of green spaces and forest reserves are traded for buildings and infrastructures; generating domestic and industrial waste in massive amounts, as urban areas are also known to host most industries, automobiles and other machines that have the capacity to emit substantial GHG gases.

As has been unveiled above, the concept of automated cities driven by IoT, AI and Big Data and other advanced technologies is foreseen to have some positive impacts in the mitigation of climate change (White et al. 2018). In particular, the ability to collect and analyze data and subsequently execute commands in real time is thought to be an effective tool toward this menace, and in areas where this has been implemented, there are positive and verifiable outcomes that have been achieved. Zvolska et al. (2019) explain that this has been made even interesting by the ability of these technologies to connect and exchange data and information among themselves.

The increased connectivity can help to respond to climate change in a more effective manner, especially due to the ability to have information shared in real-time across sectors. On this, Carter et al. (2015) highlight that connectivity allows for progressive spatial planning which in turn factors in the need for adaptability and resilience which are critical factors in combating impacts of climate change. This is made possible by the increased collaborative environment that the technologies create; more so due to availability of massive generated data from different urban fabrics and spheres. Nevertheless, as much as these technologies are providing a glimpse of hope in the fight against climate change, their installation and application still faces unprecedented challenges that would only be overcome over time (IRP 2018; Fishman and Flynn 2018; Zoonen 2016; Abas Kalair et al. 2015; Shelton et al. 2014). Okafor and Delaney (2019) highlight that such challenges include the prohibitive costs of acquiring the different technologies, their infrastructure and tools especially if such are targeted for an expansive urban area.

Additionally, the lack of formalized, and standardized architecture and a complete system that can help in handling, processing and analyzing the substantial generated data from different quarters. The technologies are also prone to other external forces like increased political interference like is the case of 5G role out that has sparked political hurdles between China -with its Huawei 5G intended roll-out in its mobile devices (Mascitelli and Chung 2019; Stokel-Walker 2019), and the United States which argue that such technologies would aid in espionage and increase infringement of privacy and security (Bond and Kynge 2018). Another hurdle that need to be overcome in the full-scale installation of these technologies is that of internal economic agendas of different cities, which have been seen to influence the installation capacity of these technologies. Such economic trends are influenced by a myriad of factors, which include the competition between different cities, as each desire to outwit and outgrow their competitors; thus, attract more business, tourists, FDIs and other such economic frontiers (Skowron and Flynn 2018; Fishman and Flynn 2018). Evans et al. (2019) underpin that economic situations in cities are also influenced by the entrepreneurialism and profit-seeking behaviors of both the public and private sector, where the public-private partnership that are established are based on such grounds. This highlights why there exists such wide gaps in the growth of different cities, where those that have firm financial grounding are seen to progress much faster, especially in the adoption of the smart technologies than their counterparts whose major source of funding are the private sector (Behzadfar et al. 2017; Woherem and Odedra-Straub 2017).

Such trends do not only hamper the procurement of advanced urban technologies but also takes a toll on the ecological conservation. The lax on the ecology further exacerbate the impacts of climate change, as these have for long been instrumental in such areas like carbon capture (Gamarra et al. 2019); hence, there need to be a sound focus on improving the urban infrastructure especially through the actualization of the autonomous cities concept, which does not only help on the urban environment, but has far reaching resulting impacts on the ecology and quality of life.

## CONCLUSION

The next evolutive phase of the Smart City concept, soon to be accelerated through the proliferation of IoT devices resulting from the introduction of the 5G network, will see the demand for automated decisions aimed at better responding to increasing the efficiency and performance of cities. While this may create governance ethical concerns relating to sensitive subjects like religion and culture, the advent of automating urban processes to respond to climate urgencies on the other hand can be incredibly useful, with positive repercussions at regional level when cities are further connected and made to exchange information between one another. One highlighted point of concern however to this actualization in the lack of standardized protocols and network on which various ICT providers operate.

## REFERENCES

Abaker, I., Hashem, T., Chang, V., & Anuar, N. B. (2016). The Role of Big Data in Smart City. *International Journal of Information Management, 36*(5), 748–758.

Abas Kalair, N., Kalair, A., & Khan, N. (2015). Review of Fossil Fuels and Future Energy Technologies. *Futures, 69,* 31–49.

Abril, D. (2018, January 11). Ericsson to Build Smart Traffic System. Retrieved from https://www.dmagazine.com/business-economy/2018/01/dallas-teams-up-with-ericsson-to-build-smart-traffic-system/

Al Nuaimi, E., Al Neyadi, H., Mohamed, N., & Al-Jaroodi, J. (2015). Applications of Big Data to Smart Cities. *Journal of Internet Services and Applications, 6*(1), 25.

Alfredsson, E., Bengtsson, M., Brwon, H. S., Isenhour, C., Lorek, S., & Stevis, D. (2018). Why Achieving the Paris Agreement Requires Reduced Overall Consumption and Production. *Sustainability: Science, Practice and Policy, 14*(1), 1–5.

Al-Fugaha, A., Guizani, M., Mehdi, M., Mohammed, A., & Moussa, A. (2015). Internet of Things: A Survey on Enabling Technologies, Protocols, and Applications. *IEEE Communications Surveys & Tutorials, 17*(4), 2347–2376.

Aliyu, A. A., & Amadu, L. (2017). Urbanization, Cities, and Health: The Challenges to Nigeria—A Review. *Annals of African Medicine, 16*(4), 149–158.

Allam, M. Z. (2018). *Redefining the Smart City: Culture, Metabolism and Governance. Case Study of Port Louis, Mauritius* (PhD), Curtin University, Perth, Australia. Retrieved from https://espace.curtin.edu.au/handle/20.500.11937/70707

Allam, Z. (2019). *Theology and Urban Sustainability.* Springer International Publishing.

Allam, Z. (2020a). *Cities and the Digital Revolution: Aligning Technology and Humanity.* Springer International Publishing.

Allam, Z. (2020b). Data as the New Driving Gears of Urbanization. In Z. Allam (Ed.), *Cities and the Digital Revolution: Aligning Technology and Humanity* (pp. 1–29). Cham: Springer International Publishing.

Allam, Z. (2020c). Digital Urban Networks and Social Media. In Z. Allam (Ed.), *Cities and the Digital Revolution: Aligning Technology and Humanity* (pp. 61–83). Cham: Springer International Publishing.

Allam, Z. (2020d). On Culture, Technology and Global Cities. In Z. Allam (Ed.), *Cities and the Digital Revolution: Aligning Technology and Humanity* (pp. 107–124). Cham: Springer International Publishing.

Allam, Z. (2020e). Privatization and Privacy in the Digital City. In Z. Allam (Ed.), *Cities and the Digital Revolution: Aligning Technology and Humanity* (pp. 85–106). Cham: Springer International Publishing.

Allam, Z. (2020f). Theology, Spirituality, and Urban Objectivity. In Z. Allam (Ed.), *Theology and Urban Sustainability* (pp. 69–79). Cham: Springer International Publishing.

Allam, Z. (2020g). Theology, Sustainability and Big Data. In Z. Allam (Ed.), *Theology and Urban Sustainability* (pp. 53–67). Cham: Springer International Publishing.

Allam, Z. (2020h). Urban and Graveyard Sprawl: The Unsustainability of Death. In Z. Allam (Ed.), *Theology and Urban Sustainability* (pp. 37–52). Cham: Springer International Publishing.

Allam, Z., & Dhunny, Z. A. (2019). On Big Data, Artificial Intelligence and Smart Cities. *Cities, 89*, 80–91.

Allam, Z., & Newman, P. (2018). Redefining the Smart City: Culture, Metabolism & Governance. *Smart Cities, 1*, 4–25.

Australian Government. (2018). Intelligent Transport Systems. Retrieved from https://www.austrade.gov.au/future-transport/intelligent-transport-systems/

Aydiner, A. S., Tatoglu, E., Bayraktar, E., Zaim, S., & Delen, D. (2019). Business Analytics and Firm Performance: The Mediating Role of Business Process Performance. *Journal of Business Research, 96*, 228–237.

Barkham, R., Bokhari, S., & Saiz, A. (2018). *Urban Big Data: City Management and Real Estate Markets.*

Bassoo, V., Ramnarain-Seetohul, V., Hurbungs, V., Fowdur, T. P., & Beeharry, Y. (2018). Big Data Analytics for Smart Cities. In N. Dey, A. Hassanien, C. Bhatt, & S. Stapathy (Eds.), *Internet of Things and Big Data Analytics Toward Next-Generation Intelligence. Studies in Big Data* (Vol. 30). Cham: Springer.

Batty, M. (2016). Big Data and the City. *Built Environment, 42*(3), 321–337.

Behzadfar, M., Ghalehnoee, M., Dadkhah, M., & Haghighi, N. M. (2017). International Challenges of Smart Cities. *Armanshahr Architecture & Urban Development, 10*(20), 79–90.

Bergman, N. (2018). Impacts of the Fossil Fuel Divestment Movement: Effects on Finance, Policy and Public Discourse. *Sustainability, 10*, 2529.

Bhadani, A. J. (2016). Big Data: Challenges, Opportunities and Realities. In M. K. Singh & D. G. Kumar (Eds.), *Effective Big Data Management and Opportunities for Implemention* (pp. 1–24). Pennsylvania, USA: IGI Global.

Bibri, S. E. (2018). The IoT for Smart Sustainable Cities of the Future: An Analytical Framework for Sensor-based Big Data Applications for Environmental Sustainability. *Sustainable Cities and Society, 38*, 230–253.

Bibri, S. E., & Krogstie, J. (2019). Generating a Vision for Smart Sustainable Cities of the Future: a Scholarly Backcasting Approach. *European Journal of Futures Research, 7*(1), 5.

Bond, D., & Kynge, J. (2018, December 3). Huawei Under Fire as Politician Fret Over 5G Security. Retrieved from https://www.ft.com/content/8bb75604-f4a6-11e8-ae55-df4bf40f9d0d

Botta, A., Donato, W. D., Persico, V., & Pescapé, A. (2016). Integration of Cloud Computing and Internet of Things: A Survey. *Future Generation Computer Systems, 56*, 684–700.

Carlos, T., Abajo, B., Feliu, E., Mendizabal, M., Fernández, J. A. M. J. G., Laburu, T., & Lejarazu, A. (2017). Profiling Urban Vulnerabilities to Climate Change: An Indicator-based Vulnerability Assessment for European Cities. *Ecological Indicators, 78*, 142–155.

Carreón, J. R., & Worrell, E. (2018). Urban Energy Systems within the Transition to Sustainable Development. A Research Agenda for Urban Metabolism. *Resource, Conservation and Recycling, 132*, 258–266.

Carter, J. G., Cavan, G., Connelly, A., Guy, S., Handley, J., & Kazmierczak, A. (2015). Climate Change and the City: Building Capacity for Urban Adaptation. *Progress in Planning, 95*, 1–66.

Chancel, L., & Piketty, T. (2015). *Carbon and Inequality: from Kyoto to Paris—Trends in the Global Inequality of Carbon Emissions (1998–2013) & Prospects for an Equitable Adaptation Fund.*

Cheng, J., Chen, W., Tao, F., & Lin, C.-L. (2018). Industrial IoT in 5G Environment Towards Smart Manufacturing. *Journal of Industrial Information Integration, 10,* 10–19.

Citron, R. (2019, August 13). Advance Traffic Management is the Next Big Thing Forsmart Cities. Retrieved from https://www.greenbiz.com/article/advanced-traffic-management-next-big-thing-smart-cities

Condoluci, M., & Mahmoodi, T. (2018). Softwarization and Virtualization in 5G Mobile Networks: Benefits, Trends and Challenges. *Computer Networks, 146,* 65–84.

Da Cruz, N. F., Rode, P., & McQuarrie, M. (2019). New Urban Governance: A Review of Current Themes and Future Priorities. *Journal of Urban Affairs, 41*(1), 1–19.

Danigelis, A. (2017, September 28). Smart City Market Growth Driven by Companies Like Schneider Electric and Ingersoll Rand. Retrieved from https://www.energymanagertoday.com/smart-city-market-growth-driven-companies-like-schneider-electric-ingersoll-rand-0171629/

Diffenbaugh, N. S., & Burke, M. (2019). Global Warming has Increased Global Economic Inequality. *PNAS, 116*(20), 9808–9813.

Drell, L. (2011). 4 Cities Using Tech to Alleviate Traffic. Retrieved from https://mashable.com/2011/11/16/traffic-tech/

Dutta Pramanik, P., Pal, S., & Choudhury, P. (2018). Beyond Automation: The Cognitive IoT. Artificial Intelligence Brings Sense to the Internet of Things. In A. K. Sangaiah, A. Thangavelu, & V. Meenakshi Sundaram (Eds.), *Cognitive Computing for Big Data Systems Over IoT: Frameworks, Tools and Applications* (pp. 1–37). Cham, Switzerland: Springer International Publishing.

Erler, M. (2019, April 11). Playing Intelligence. Retrieved from https://jods.mitpress.mit.edu/pub/0l8x7kip

Ersoy, A., & Alberto, K. C. (2019). Understanding Urban Infrastructure via Big Data: The Case of Belo Horizonte. *Regional Studies, Regional Science, 6*(1), 374–379.

Evans, D. (2011, April). *How the Next Evolution of the internet is Changing Everything.* The Internet of Things (White Paper). Cisco Internet Business Solutions Group.

Evans, J., Karvonen, A., Luque-Ayala, A., Martin, C., McCormick, K., Raven, R., & Palgan, Y. V. (2019). Smart and Sustainable Cities? Pipedreams, Practicalities and Possibilities. *Local Environment, 24*(7), 557–564.

EY. (2014). *Big Data: Changing the Way Businesses Compete and Operate.* Retrieved from https://www.ey.com/Publication/vwLUAssets/EY_-_Big_

data:_changing_the_way_businesses_operate/%24FILE/EY-Insights-on-GRC-Big-data.pdf

Fan, J.-L., Kong, L.-S., Wang, H., & Zhang, X. (2019). A Water-energy Nexus Review from the Perspective of Urban Metabolism. *Ecological Modelling, 392*, 128–136.

Ferrández-Pastor, F.-J., Gómez-Trillo, S., Nieto-Hidalgo, M., García-Chamizo, J.-M., & Valdivieso-Sarabia, R. (2018). Intelligent Power Management System Using Hybrid Renewable Energy Resources and Decision Tree Approach. *Proceedings, 2*(19), 1239.

Ferrero Bermejo, J., Gómez Fernández, J. F., Olivencia Polo, F., & Crespo Márquez, A. (2019). A Review of the Use of Artificial Neural Network Models for Energy and Reliability Prediction. A Study of the Solar PV, Hydraulic and Wind Energy Sources. *Applied Sciences, 9*(9), 1844.

Fisher, T. (2019). 5G Availability Around the World. Retrieved from https://www.lifewire.com/5g-availability-world-4156244

Fishman, T. D., & Flynn, M. (2018). Part Two: Funding and Financing Smart Cities Series. *Using Public-Private Partnerships to Advance Smart Cities*, 1–10.

Freudendal-Pedersen, M., Kesselring, S., & Servou, E. (2019). What is Smart for the Future City? Mobilities and Automation. *Sustainability, 11*(1), 221.

Gamarra, N. C., Correia, R. A., Bragagnolo, C., Campos-Silva, J. V., Jepson, P. R., Ladle, R. J., & Mendes Malhado, A. C. (2019). Are Protected Areas undervalued? An Asset-based Analysis of Brazilian Protected Area Management Plans. *Journal of Environmental Management, 249*, 109347.

Garcia, A. (2019, April 9). Looking for 5G? Here are the US Cities that Have It. Retrieved from https://edition.cnn.com/2019/04/09/tech/5g-network-us-cities/index.html

Gil, D., Ferrández, A., Moramora, H., & Peral, J. (2016). Internet of Things: A Review of Surveys Based on Context Aware Intelligent Services. *Sensors, 16*, 1069.

Goldstein, P. (2019). How Will 5G Networks Impact Smart Cities? Retrieved from https://statetechmagazine.com/article/2019/08/how-will-5g-networks-impact-smart-cities-perfcon

Grand View Research. (2019, May). Smart Cities Market Size Worth $237.6 Billion by 2025 | CARG: 18.9%. Retrieved from https://www.grandviewresearch.com/press-release/global-smart-cities-market

Guo, K., Lu, Y., Gao, H., & Cao, R. (2018). Artificial Intelligence-Based Semantic Internet of Things in a User-Centric Smart City. *Sensors, 18*(5), 1341–1363.

Huikkola, T., & Kohtamäki, M. (2019). Interplay of Strategic Orientations in the Development of Smart Solutions. *Procedia CIRP, 83*, 89–94.

IRENA. (2018). *Renewable Energy Market Analysis: Southeast Asia*. Retrieved from Abu Dhabi https://www.irena.org/-/media/Files/IRENA/Agency/Publication/2018/Jan/IRENA_Market_Southeast_Asia_2018.pdf

IRP. (2018). *The Weight of Cities: Resource Requirements of Future Urbanization*.

Ishida, T. (2017). *Digitial City, Smart City and Beyond*. Paper Presented at the 26th World Wide Web Intenational conference (WWW17), Perth, Australia.

ITU. (2018). *Assessing the Economic Impact of Artificial Intelligence*.

Jarrahi, M. H. (2018). Artificial Intelligence and the Future of Work: Human-AI Symbiosis in Organizational Decision Making. *Business Horizons, 61*, 577–586.

Joss, S., Sengers, F., Schraven, D., Caprotti, F., & Dayot, Y. (2019). The Smart City as Global Discourse: Storylines and Critical Junctures Across 27 Cities. *Journal of Urban Technology, 26*(1), 3–34.

Kamel Boulos, M. N., Tsouros, A. D., & Holopainen, A. (2015). 'Social, Innovative and Smart Cities are Happy and Resilient': Insights from the WHO EURO 2014 International Healthy Cities Conference. *International Journal of Health Geographics, 14*(1), 3.

Kolotouchkina, O., & Seisdedos, G. (2017). Place Branding Strategies in the Context of New Smart Cities: Songdo IBD, Masdar and Skolkovo. *Place Branding and Public Diplomacy, 14*(2), 115–124.

Kylili, A., & Paris, A. F. (2015). European Smart Cities: The Role of Zero Energy Buildings. *Sustainable Cities and Society, 15*, 86–95.

Lim, C., Kim, K.-J., & Maglio, P. P. (2018). Smart Cities with Big Data: Reference Models, Challenges, and Considerations. *Cities, 82*, 86–99.

Lomas, N. (2019). UK's First 5G Network Taster goes Live in Six Cities Tomorrow. Retrieved from https://techcrunch.com/2019/05/29/uks-first-5g-network-taster-goes-live-in-six-cities-tomorrow/

Loucks, J., & Hupfer, S. (2018). Companies Boost AI Spending, But Risks, Talent Gaps Persist. Retrieved from https://deloitte.wsj.com/riskandcompliance/2019/09/25/companies-boost-ai-spending-but-risks-talent-gaps-persist/

Lozada, N., Arias-Pérez, J., & Perdomo-Charry, G. (2019). Big Data Analytics Capability and co-Innovation: An Empirical Study. *Heliyon, 5*(10), e02541.

Ma, S., Chen, X., Li, Z., & Yang, Y. (2018). A Retrival Optimized Surveillance Video Storage System for Campus Application Scenarios. *Journal of Electrical and Computer Engineering*, 10. Retrieved from Hindawi website: https://doi.org/10.1155/2018/3839104

Mahdavinejad, M. S., Rezvan, M., Barekatain, M., Adabi, P., Barnaghi, P., & Sheth, A. P. (2018). Machine Learning for Internet of Things Data Analysis: A Survey. *Digital Communications and Networks, 4*(3), 161–175.

Marzano, G., Lizut, J., & Siguencia, L. O. (2019). Crowdsourcing Solutions for Supporting Urban Mobility. *Procedia Computer Science, 149*, 542–547.

Mascitelli, B., & Chung, M. (2019). Hue and Cry Over Huawei: Cold War Tensions, Security Threats or Anti-competitive Behaviour? *Research in Globalization, 1*, 100002.

McKinsey & Company. (2018). *Smart City Solutions: What Drives Citizen Adoption around the Globe?*

Minoli, D., Sohraby, K., & Occhiogrosso, B. (2017). IoT Considerations, Requirements, and Architectures for Smart Buildings—Energy Optimization and Next-Generation Building Management Systems. *IEEE Internet of Things Journal, 4*(1), 269–283.

Naganathan, V., & Rao, R. K. (2018). The Evolution of Internet of Things: Bringing the power of Artificial Intelligence to IoT, its Opportunities and Challenges. *International Journal of Computer Science Trends and Technology, 6*(3), 94–108.

Newman, P. (2010). Green Urbanism and its Application to Singapore. *Environment and Urbanization ASIA, 1*(2), 149–170.

Noura, M., Atiquzzaman, M., & Gaedke, M. J. M. N. (2019). Interoperability in Internet of Things: Taxonomies and Open Challenges. *Mobile Networks and Applications, 24*(3), 796–809.

Novelli, M. (2005). *Niche Tourism: Contemporary Issues, Trends and Cases.* Burlington, MA: Elsevier Butterworth-Heinemann.

Obinikpo, A. A., & Kantarci, B. (2017). Big Sensed Data Meets Deep Learning for Smarter Health Care in Smart Cities. *Journal of Sensor and Actuator Networks, 6*, 1–26.

Ochola, G. O. (2018). Urbanization and Environmental Stress: A Review of Impacts of Urban Development on the Environment in Kenya. *International Journal of Environmental Science & Natural Resources, 14*(4), 1–5.

Oh, D.-S., & Phillips, F. (2014). *Technopolis: Best Practices for Science and Technology Cities.* London: Springer-Verlag.

Okafor, N. U., & Delaney, D. (2019). Considerations for System Design in IoT-Based Autonomous Ecological Sensing. *Procedia Computer Science, 155*, 258–267.

Ortiz, L., Gonzalez, J., & Lin, W. (2018). Climate Change Impacts on Peak Building Cooling Energy Demand in a Coastal Megacity. *Environment Research Letters, 13*(9), 1–10.

Papavasiliou, A., & Oren, S. S. (2014). Large-Scale Integration of Deferrable Demand and Renewable Energy Sources. *IEEE Power and Energy Magazine, 29*, 489–499.

Park, E., Del Pobil, A. P., & Kwon, S. J. (2018). The Role of Internet of Things (IoT) in Smart Cities: Technology Roadmap-oriented Approaches. *Sustainability, 10*(5), 1388.

Patel, K. K., & Patel, S. M. (2016). Internet of Things-IoT: Definition, Characteristics, Architecture, Enabling Technologies, Applications and Future Challenges. *International Journal of Engineering Science and Computing, 6*(5), 6122–6131.

Payne, K. (2018). Artificial Intelligence: A Revolution in Strategic Affairs? *Survival, 60*(5), 7–32.

Popovič, A., Hackney, R., Tassabehji, R., & Castelli, M. J. I. S. F. (2018). The Impact of Big Data Analytics on firms' High Value Business Performance. *Information Systems Frontiers, 20*(2), 209–222.

Purdy, M., & Daugherty, P. (2017). *How AI Boosts Industrial Profits and Innovation*. Retrieved from https://www.accenture.com/fr-fr/_acnmedia/36 dc7f76eab444cab6a7f44017cc3997.pdf

Purkus, A., Gawel, E., & Thrän, D. (2017). Addressing Uncertainty in Decarbonisation Policy Mixes—Lessons Learned from German and European Bioenergy Policy. *Energy Research & Social Science, 33*, 82–94.

PWC. (2018). *Macroeconomic Impact of Artificial Intelligence*. Retrieved from https://www.pwc.co.uk/economic-services/assets/macroeconomic-impact-of-ai-technical-report-feb-18.pdf

REN21. (2018). *Renewables 2018 Global Status Report*.

Richards, D. R., & Thompson, B. S. (2019). Urban Ecosystems: A New Frontier for Payments for Ecosystem Services. *People and Nature, 1*(2), 249–261.

Richter, C., Kraus, S., & Syrjä, P. (2015). The Smart City as an Opportunity for Entrepreneurship. *International Journal of Enterpreneurial Venturing, 7*(3), 211–226.

Rishi, R., & Saluja, R. (2019). Future of IoT. Retrieved from Ernest and Young website https://www.ey.com/Publication/vwLUAssets/EY_-_Future_of_IoT/$FILE/EY-future-of-lot.pdf

Salingaros, N. A. (2000). Complexity and Urban Coherence. *Journal of Urban Design, 5*, 291–316.

Salingaros, N. A. (2014). Complexity in Architecture and Design. *Oz Journal, 36*, 18–25.

Scarcello, F. (2018). Artificial Intelligence. *Reference Module in Life Sciences*, 1–7. Retrieved from https://www.researchgate.net/publication/323220519_Artificial_Intelligence

Shelton, T., Zook, M., & Wiig, A. (2014). The "actually existing smart city". *Cambridge Journal of Regions, Economy and Society, 8*(1), 13–25.

Shepard, W. (2017, December 12). Why Hundreds of Completely New Cities are Being Built around the World. Retrieved from https://www.forbes.com/sites/wadeshepard/2017/12/12/why-hundreds-of-completely-new-cities-are-being-built-around-the-world/#1380073e14bf

Silva, B. N., Khan, M., & Han, K. (2018). Towards Sustainable Smart Cities: A Review of Trends, Architectures, Components, and Open Challenges in Smart Cities. *Sustainable Cities and Society, 38*, 697–713.

Singh, S. (2019). *Smart Cities Market Worth $717.2 billion by 2023*.

Skowron, J., & Flynn, M. (2018). *The Challenge of Paying for Smart Cities Projects*.

Slavova, M., & Okwechime, E. (2016). African Smart Cities Strategies for Agenda 2063. *Africa Journal of Management, 2*(2), 210–229.

Snow, J. (2017, July 20). This AI Traffic System in Pittsburgh has Reduced Travel Time by 25%. Retrieved from https://www.smartcitiesdive.com/news/this-ai-traffic-system-in-pittsburgh-has-reduced-travel-time-by-25/447494/

Stokel-Walker, C. (2019). Banning Huawei from 5G Infrastructure. *New Scientist, 242*(3228), 11.

Taecharungroj, V., Muthuta, M., & Boonchaiyapruek, P. (2019). Sustainability as a Place Brand Position: A Resident-centric Analysis of the Ten Towns in the Vicinity of Bangkok. *Place Branding and Public Diplomacy, 15*, 1–19.

Taeihagh, A., & Lim, H. S. M. (2019). Governing Autonomous Vehicles: Emerging Responses for Safety, Liability, Privacy, Cybersecurity, and Industry Risks. *Transport Reviews, 39*(1), 103–128.

Taylor Buck, N., & While, A. (2015). Competitive Urbanism and the Limits to Smart City Innovation: The UK Future Cities initiative. *Urban Studies, 54*(2), 501–519.

Taylor Buck, N., & While, A. (2017). Competitive Urbanism and the Limits to Smart City Innovation: The UK Future Cities Initiative. *Urban Studies Journal Limited, 54*(2), 501–519.

Tu, Y. (2017). Urban Debates for Climate Change after the Kyoto Protocol. *Urban Studies, 55*(1), 3–18.

Tzafestas, S. G. (2018). Synergy of IoT and AI in Modern Society: The Robotics and Automation Case. *Robotics & Automation Engineering Journal, 31*(5), 1–15.

UN Environment. (2019). Cities and Climate Change. Retrieved from https://www.unenvironment.org/explore-topics/resource-efficiency/what-we-do/cities/cities-and-climate-change

United Nations. (2016, May 9–13). *Smart Cities and Infrastructure: Report of the Secretary-General.* Paper Presented at the Commission on Science and Technology for Development, Geneva.

United Nations. (2019). Shaping our Future Together. Retrieved from https://www.un.org/en/sections/issues-depth/climate-change/index.html

Van Winden, W., & van den Buuse, D. (2017). Smart City Pilot Projects: Exploring the Dimensions and Conditions of Scaling Up. *Journal of Urban Technology, 24*(4), 51–72.

White, M., Margolies, J., Ronanki, R., Steier, D., Tuff, G., Bhattacharya, A. ... Saif, I. (2018). Exponential Technology Watch List: Innovation Opportunities on the Horizon. In D. Insight (Ed.), *Tech Trends: The Symphonic Enterprise*: Deloitte Insight.

Woherem, E. E., & Odedra-Straub, M. (2017). The Potentials and Challenges of Developing Smart Cities in Africa. *Circulation in Computer Science, 2*(4), 27–39.

Yavuz, M. C., Cavusoglu, M., & Corbaci, A. (2018). Reinventing Tourism Cities: Examining Technologies, Applications and City Branding in Leading Smart

Cities. *International Interdisciplinary Business-Economics Advancement Journal, 3*(1), 57–70.

Yigitcanlar, T., & Bulu, M. (2016). Urban Knowledge and Innovation Spaces. *Journal of Urban Technology, 23*(1), 1–9.

Zanella, A., Bui, N., Castellani, A., Vangelista, L., & Zorzi, M. (2014). Internet of Things for Smart Cities. *IEEE Internet of Things Journal, 1*(1), 22–32.

Zhou, K., Fu, C., & Yang, S. (2016). Big Data Driven Smart Energy Management: From Big Data to Big Insights. *Renewable and Sustainable Energy Reviews, 56*, 215–225.

Zoonen, L. v. (2016). Privacy Concerns in Smart Cities. *Government Information Quarterly, 33*(3), 472–480.

Zorins, A., & Grabusts, P. (2015). *Artificial Neural Networks and Human Brain: Survey of Improvement Possibilities of Learning.* Paper Presented at the 10th International Scientific and Practical Conference, Rezekne, Latvia.

Zvolska, L., Lehner, M., Voytenko Palgan, Y., Mont, O., & Plepys, A. (2019). Urban Sharing in Smart Cities: The Cases of Berlin and London. *Local Environment, 24*(7), 628–645.

# CHAPTER 3

# On Complexity, Connectivity and Autonomy in Future Cities

**Abstract** The smart cities concept is gaining in popularity and as the advent of technology is cities is being enforced by both policies and private corporations, the number of Internet of Things (IoT) devices in cities is soaring. This is leading to a complex array of devices that are made to communicate seamless between each other leading to an additional, and complex, digital layer that can assist with creating a more intelligent urban fabric; leading closer to the actualization of the autonomous cities concept. Ideas on achieving complexity in cities is however not new, but there has been little progress on this end in regenerative policies and projects in modernist new towns. This chapter showcases how the advent of technology can help in its actualization and in the process lead to more inclusive, safe, sustainable and resilient cities, as underlined in the Sustainable Development Goal 11.

**Keywords** Complexity • Autonomous cities • Connectivity • 5G • Internet of Things (IoT) • Future cities

© The Author(s) 2021
Z. Allam, *The Rise of Autonomous Smart Cities*, Sustainable Urban Futures, https://doi.org/10.1007/978-3-030-59448-0_3

# INTRODUCTION

The smart cities concept is gaining popularity all over the world, following its promise to solve some of the perennial challenges that global cities are experiencing in this twenty-first century. The pinnacle of smart cities that makes the concept a global buzzword is its increased reliance on information technology and capitalizing on advanced technologies in different aspects of the urban fabric (Allam 2020b, c). Ishida (2017) notes that a majority of the notable cities that have earned the accolade of being 'smart' have integrated IT services in areas like spatial planning, service delivery, data collection and sharing and monitoring and controlling different urban fabrics. The results of trusting IT, as noted in a wide array of literature include, attractiveness of cities that has been instrumental in attracting a wide spectrum of economic activities like foreign direct investments (FDIs), and tourism (local and international tourism) (Allam and Jones 2019a; Allam and Newman 2018). It is also hailed for increased efficiency in service delivery and real-time response to different issues, more so by moving beyond IT and integrating it to real issues like sustainability, public participation and social inclusivity. On this, Caird and Hallett (2018) also associate the concept of smart cities with novel strategies of dealing with urbanization, population increase and climate change, which are synonymous with most global cities.

The promise of unprecedented growth and development, among other positives associated with smart cities concepts, is said to be the magnet that is pulling global cities to turning to this concept. In perspective, today, as noted by Joss et al. (2019), there over 27 cities that satisfy all the requirements to earn the classification as being a 'smart city'. These were identified from a list of 5,553 global cities that have a population of 100,000 and above. The identification process was done following a systematic webometric exercise that was conducted on all the said cities. Another report by *Eden Strategy Institute and ONG&ONG Pte* (2018) showcases that the world has approximately 50 cities that can be termed as 'smart' especially in respect to how they are governed. This report identifies issues like planning, implementation and funding of smart cities initiatives. It also captures how the identified cities are in the forefront in ensuring digital inclusion, they have open data and are doing everything possible to prepare their workforce for smart initiatives as well as embracing smart cities leadership models.

Besides the city managers, the concept of smart cities is also seen to be driven by other stakeholders like ICT Corporations, which have been very active in pushing technologically inclined products with the narrative that those can increase the efficiency and performance of cities (Danigelis 2017). Most of these products are customized such that they are able to perform under the Internet of Things (IoT) platform, which has proven critical in support of the concept of digital transformation and revenue creation opportunities. For this reason, it has been observed that the number of IoT devices in the market is rising, with large ICT Corporations leading in their production and supply of these products hosting cutting edge technologies. A report by Rishi and Saluja (2019) of EY supports the claim of increasing devices by highlighting that their numbers has increased from only 7 billion in 2017 to an expected surplus of 25 billion devices by 2025. With such large numbers, the revenue accrued from their market is expected to exceed USD 1.1 trillion. A different market research done by Statista Research Department (2016) projects that the number of connected IoT devices will reach 75.44 billion devices globally by 2025; namely a 489.55% increase from the 15.41 billion devices that were recorded in 2015. A different report by ICT giants, Ericsson (2019) predicts that by 2022, the number of connected devices will be approximately 29 billion.

On market value, just like EY predicted, Statista Research Department predicts that by the end of this year (2019), the market value will reach a high of USD 212 billion, an exponential increase from the 100 billion that was recorded in 2017. Going forward, by 2025, they predict that the market value will reach a high of USD 1.6 trillion (Liu 2019). The sheer size and profitability of this market, if such figures are accurately represented, are serving as the ground by which ICT corporations are investing heavily in R&D to ensure they control as much market as possible. These have also been credited for the increased partnership between large high-tech companies and local governments in providing digital solutions in cities.

However, while such devices are critical in the actualization of the smart cities concept, it has been argued that they need to be integrated with other tools to help achieve the objective of sustainability and inclusivity in urban fabrics that is anticipated in the Sustainable Development Goals (SDGs). From a report by IoT Analytics (Lueth and Pasqua 2019), it was shown that the available IoT deployments has already integrated and helped actualize a number of SDGs which include SDG 11 on Smart Cities and Communication, SDG9 on Industry, innovation and

Infrastructure, and SDG 3 on Good health and well-being. Others include SDG12 on Responsible production and consumption and SDG7 on Affordable and clean energy. These examples show that IoT has the potential to allow the attainment of all the SDGs, especially in respect to monitoring and controlling on the unconnected devices which make the 'complexity theory' on urban systems even more interesting (Allam 2019b, 2020d; Allam and Dhunny 2019). These are even more attainable when placed in the hands of the private sector, which Arias et al. (2018) opine control 84% of all the IoT deployments with 75% of this directly involved in actualizing the said SDGs.

The emergence and acceptance of technologies in cities, especially via the smart cities initiative, can thus create exciting new prospects leading to perhaps surprising support from proponents of the New Urbanism concept. Arias et al. (2018) acknowledge that the concept of smart cities has the potential to spark and inject new thinking not only in spatial planning but also in areas like governance, sustainability, socialism, economic growth and environment. This is possible as those concepts are seen to support interconnectedness on all aspects, thus, making cities more interactive, with both top-down and bottom-up forms of communication and interactions being even more feasible (Manyika et al. 2015), as will be discussed in the succeeding sections.

## COMPLEXITY THEORY AND IoT DEVICES

From the ancient days, cities have always served as the foundational structure hosting different human activities and contributing greatly in the growth of the national and/or regional economies. Such have been possible due to the complex nature that characterized many cities as noted by different notable urban theorists like Jane Jacobs (Jacobs 1961), Christopher Alexander (Alexander 1965, 1979, 2002; Alexander et al. 1977) and Nikos Salingaros (Salingaros 1998, 2014) amongst many others. On this, it has been observed that they advocate a common principle of rendering cities that are intricately connected such that those living therein have more opportunities to interact, experience richer quality of life and enjoy peace and tranquility that is premised on sound, and well thought urban fabrics. The concept of interconnectedness was well articulated in traditional cities, where streets and living blocks, for example, were well connected by well designed.

On the contrary, in modern days, modernism is seen to negate most of the traditional principles and are now advocating for linear and centralized processes that are seen to render issues of inequalities even more profound. In the words of Salingaros (Salingaros 2000, 2003, 2006) and Alexander (Alexander 2002), such cities are now seen to as mechanical cities that are void of life, unhappiness, and do not promote 'wholeness'. On such, Jane Jacobs (Jacobs 1961) was always against, especially on what she argued are aimed at decentralizing the residences through the separation of cities into commercial, industrial residential and parks section. In her observation, high-rise buildings that were proposed in America during the 1960s aimed at replacing old buildings would actualize such separations, and in the process, would increase the social separation that could not augur well with urban liveability. The solution to such, which is still valid to date is the adoption of mixed-use approach where new buildings would accommodate a diverse use—including commercial and residential.

Like Jane Jacobs, Salingaros (2000) is seen to critique modernism and the way it is seen to override traditional urban planning. In particular, with the rising demands for such things like 'green' architecture brought about by urbanization, population growth and climate change, Salingaros does not think that modernism has the solutions to respond to the challenges of the day. According to him, modernism promotes architectural practices that are not conducive for humanity, and which do not invoke the sense of belonging or harmony. Another author, Mehaffy (expounds that even in the modernist realm, the intricate and ingenuity of traditional architectural practices still remain relevant, such that, even when urbanists or architects want to deviate from such, they end up having 'incoherent white noise of disordered structures'. This, he notes, is despite the availability of advanced technologies that are seen to greatly influence modernist thinking of architectural and urban structures. To him, like his other counterparts cited above, our cities need to a mixture of modern thinking and traditional practices. Otherwise, the glamour and attractiveness being pursued will only remain to be seen as 'entertainment machines' that unfortunately are incoherent and constantly disharmonious with each other, other than maintaining the 'pattern of order', as advanced by Alexander (2002).

One wave that modernists seem to ride on while designing and actualizing their work is that of technological advancement; more so the smart cities concept. On this, they believed that with advent of IoT, some of the negatives associated with modernism are bound to be solved. With IoT, as noted previously, there has been an increase in smart devices and a wide array of sensors that have made availability of data ubiquitous in different urban domains. Such, as shared by Qin et al. (2016), have allowed cities to generate massive data from objects that were once passive, and now transformed to dynamic objects. These are interconnected into a web that make up a proactive urban system that has the capacity to generate real-time solutions to a wide range of issues after their data is processed and analyzed (Keenan 2018; Bačić et al. 2018; Kim et al. 2017). This interconnectedness in the urban fabric provide a richer understanding of cities; which is a positive prerequisite for higher complexity. Berntzen et al. (2016) highlights that this complexity is manifested when different subsystems in cities are unified into a singular system that is characterized of 'smart' (economy, people, governance, mobility, environment and living). This smartness, as Bassoo et al. (2018) note, is enabled by the quality of data generated by different urban domains and how such are processed. At this juncture, IoT is also seen to play significant role in allowing other advanced technologies like Artificial Intelligence (AI) to be incorporated in urban systems. In particular, these have been seen to assist IoT devices and sensors to become even more complex and efficient in capturing data, and also in their energy efficiency. Integration of IoT and AI, as noted by Huntingford et al. (2019), increase the potential generated data to predict future occurrences of some events like weather, traffic and resource consumption, thus, help in making prior plans to overcome such. This has been noted to be of subtle importance in areas like spatial planning and resource allocation, distribution and consumption (Allam 2019a, 2020a).

To achieve the said higher complexity, data generated by different domains need to be aggregated and made freely accessible so as to provide equal opportunities for all to tap into this rich resource (Allam et al. 2019). This is one area where modernists have failed. It has been observed that despite the number of IoT devices and sensors increasing, the issue of collaboration between different market players have been very limited, and this has resulted into existence of devices that are incompatible and requiring different networks since their functionalities and protocols are different. This has made it impossible to have a unified database and has given advantage to some players who have the capacity to manufacture and install advanced technologies. The hurdles in data accessibility arising from such has in turn sparked issues like inconsistent developments in different urban domains, lack of trust from the citizen and creation of structures that do not add up to what Alexander anticipated to be a 'pattern language' (Alexander et al. 1977).

## On Complexity and Urban Regeneration

Availability of massive data is being taunted as the 'New Oil' of this century, and probably of the future. Whether that statement is true or otherwise, it is of no doubt its access and control is disruptive especially in urban areas. With data, as is appreciated above, cities position in economies are becoming even more pronounced, as people are able to gain a deeper understanding of its functioning. Such understanding has in turn led to better governance dimensions, more so from informed decisions accentuated by availability of data on different aspects. Kamilaris and Ostermann (2018) credit the availability of data from different IoT devices with improved spatial planning in different parts of the globe, where it has been found that analysis and interpretation of such data is providing a real-time virtual 'picture' of the city, and even more, is used to predict the future. This has been argued to be critically important as urban managers can then strategize on how to overcoming the looming challenges, and those that are anticipated to happen going forward. Tomor et al. (2019) argue that such strategies have been seen to include people, in a collaborative fashion to address urban challenges; especially ones related to sustainability, urban security, resource use and information sharing. Such collaboration in turn has been seen to bring a transformation of governance from the traditional form to smart governance that is said to enable among other things, error-free, appropriate and cost effective service delivery to the citizenly of the city in a timely manner (Sarker et al. 2018).

Besides governance, the availability of big data in cities is also seen to lead to a culture of efficiency in resource management and reducing wastage, both in public and private sectors, and by so doing, there is surplus revenue to undertake urban renewal programs. One area that have greatly benefited from the advent of data is the energy sector, which has been 'thorny' in many cities, especially due to its contribution to emissions, which have even captured the attention of global audience; prompting calls for adoption of alternative energy sources. Zhang and Qiu (2018) express how the availability of big data in cities have allowed for the optimization in all stages of energy production, including leading to smart grids that are 'fed' from green energy sources (Farhangi 2010; Gellings et al. 2004). In addition, this has allowed even for small scale production of energy, especially in rooftops, and through these, those engaging in the practice are said to earn some income through the sale of surplus energy through the Peer-to-Peer platforms. Zhang and Qiu (2018) share that through data, it is possible for local government and utility companies to predict energy production, demand, distribution, and consumption among other variables, and such can allow for such practices like decentralization of energy production among other things. Besides energy, data is also instrumental in optimization other resources like water, and construction materials, and this leads to limited wastage generation. Zang and Ye (2015) share that through data, it is possible to effectively manage their human resource; thus, reducing incidences of ghost workers and leading to a leaner workforce that is easier to manage, remunerate and train for improved outputs and growth. Such is important to local governments that are characterized of huge workforces, which unfortunately does not perform at levels commensurate to their numbers.

The benefits brought about by availability of big data need to be spread across the board including cities that have for long been degenerating. That is, they need to be felt and enjoyed by all the players in equitable ways, such that each can maximize their contribution. To put this into perspective, within an urban setup, the myriad of data generated by arrays of sensors from different domains need to be made accessible such that even smaller local companies can also manage to share in the growth and development of the city. This point is raised since it has been found out that, most often, only larger ICT corporations -with the means and capacity to conduct comprehensive R&D and undertake implementation responsibilities of smart cities projects, usually have an upper hand when it comes to data handling. This is warranted by their positioning that allows them to have systems and networks that provide for data collection, storage, processing and sharing. While this is the case, new entrants usually struggle to keep up with the pace, and due to the competitive edge that the large corporations have, smaller ones end up being edged-out. This is the case even if smaller local companies and new entrants offer very solid digital solutions in various areas, besides contributing to job creation and economic growth. For instance, in the transportation sector, besides companies like Uber that have global presence, there are numerous start-ups in different cities that offer competing services at even affordable rates. Such include *Lyft* based in San Francisco, *Didi Chuxing* based in China, *OlaCabs* which is popular in Bangalore, India (RideGuru 2019), and *Little* based in Kenya (Team Little 2019) among many others found in various countries. Such have created jobs, and sparked an array of digitally powered solutions that are seen to lead to a more progressive sector, and has brought life to some cities that were somehow inaccessible. On the same sector, accessibility of data allows for services like biking that not only ease pressure on traffic, but also contributes to environmental sustainability and in the health wellbeing of those cycling.

Allowing a fair playground for all participants in urban areas regardless of their market share is seen to continue sparking an array of digitally powered solutions that benefits urban dwellers and in extension, render an urban regenerative processes. Townsend (2013) shares how the availability of data on mobile usage has sparked numerous, innovative platforms like money transfer, online shopping and other numerous Small and Medium-size Enterprises (SMEs) in different cities. Such have sparked regenerative processes even in smaller cities, as more and more businesses are set up. On this aspect, availability of data has also allowed for novel ways of preserving and sharing cultural heritage and through such initiatives, there are numerous urban areas that have been regenerated especially through such activities like tourism and emergence of new art. On this, Sasaki (2010) give an example of Osaka City, which, though having creative city policies, failed to benefit, but, after rethinking their strategy, and included the comprehensive urban strategy, their fortune started to increase. The author highlight that this was inspired by engaging the locals, and accommodating their creative diversity. Such inclusivity and collaborative participation are prerequisite for regenerative programs in cities. Another city that gives a glimpse of how data can result to regeneration is the city of Marseille, which is said to capitalize on data to foster its rich cultural heritage to foster its glory back after declining when its role as a port city started to dwindle. With data, as noted by Hickins (2017), it is heading toward being among the safest city, with prosperous accolades like a soccer city. These among many other initiatives that are enabled by availability of data in urban areas have not only brought increased economic activities, this has also led to progress in the achievement of SDG 11 and promoted inclusivity.

## PEER-TO-PEER URBANISM AND SELF MANAGEMENT
## AND CONTROL

The advent of IoT, as noted above has unlimited potentials, and as is demonstrated in this section, it is proving to be an essential platform for supporting the concept of Peer-to-peer Urbanism. From a historical perspective, the concept of P2P was initiated when people from similar backgrounds and fields started 'finding' each other via different communication platforms, especially on the internet, where they could converse and share information touching on their backgrounds. Mostly, in the initial stages, P2P was common with manufacturers, software developers and others who shared knowledge and information without the limitation of geographical boundaries or physical proximity. Following these footsteps, it is said that planners and urban designers who were working independently, adopted the concept of P2P to collaborate on different issues appertaining to urban environment, and through such efforts, P2P urbanism was born. Salingaros and Quintero (2010) note that unlike other fields where those engaged in P2P are majorly drawn from similar or related backgrounds, in P2P urbanism, the participants or collaborators are drawn from different backgrounds since the urban environment, especially in the twenty-first century entails and encompasses numerous issues, and benefits from knowledge drawn from different, and wide range of backgrounds. This is further supported today by the availability of data processing technologies like big data (Kamrowska-Zaluska and Obracht-Prondzyńska 2018; Thakuriah et al. 2017; Mjimba and Sibanda 2019; Souza et al. 2016), crowd sourcing (Marzano et al. 2019) and AI (Blanco et al. 2018; Bini 2018) among others that integrates well with IoT enabled sensors. When such data are collated, and processed together, they yield valuable information that has enabled concepts like smart cities and biophilia to be actualized in different cities around the globe.

Salingaros (2011) shares that through P2P platforms urban areas are seen to benefit in various means; where this could lead to democratization of data as these can be sourced from everyone who have access to the network. According to him, through this sharing, the governance of cities is seen to shift from the rigid top-down approach that had been imposed by governments, to a more open and collaborative bottom-up approach where citizens can flag wants and participate in how proposals and solutions should be implemented in their city. By engaging the residents, chances of creating dysfunctional urban fabric—highly criticized by Jane Jacobs, Alexander Christopher and Nikos Salingaros would then seldom happen. Where such were built, Salingaros (2011) shares that it would be easier and quicker to re-build. Same case is true with the subject of regeneration of cities, and in this case, it is argued that it could be done in such a way that there would be preservation of culture and heritage that locals, who are also part of the regeneration program, holds and this would be paramount in ensuring higher levels of liveability. These progressive achievements are somehow tied to the amount of data that is received from different quarters, and more importantly, how such data is handled and shared with the stakeholders. On this, Amoretti and Zanichelli (2018) argue that the success of a project under the P2P platform is determined by the consistency of stored data, its access and also its security; which could be actualized through the emerging concept of Blockchain technology (Allam 2018; Allam and Jones 2019b; Shahab and Allam 2020).

In our cities, where the concept of smart cities is taking shape, there are a number of hurdles that requires to be cleared for true P2P to take shape. One of this, as has been insistently indicated (Kamilaris and Ostermann 2018), is the democratization of data from larger ICT corporations, and ensuring its accessibility even by smaller, local companies and start-ups, which have been shown to play significant role in bettering the lives of urban residents. Allowing access of data to all is what Salingaros (2010) calls P2P ethics, which he perceives to be a prerequisite condition to achieving what Christopher Alexander anticipated in his 1977 book on pattern languages. According to Salingaros (2010), it is impossible to achieve or maintain the intended pattern languages unless everyone has access to the design and other information of the environment that is being built. Likewise, even in actualizing the smart cities concepts, it would be impractical for smaller companies and even residents to participate if data collection, storage, processing, dissemination and security is left in the hands of third-party companies. The alternative to such rigid approaches to handling urban data, which, in essence is meant for public consumption is to adopt and promote the concept of open source knowledge, where the urban data is made freely available to the public.

Barbosa et al. (2014) share that opening the data to the public fosters transparency and encourage collaboration between different stakeholders, and this is seen to have positive impacts in the quality of urban environment that is built from such data. The authors share that through the open urban data, numerous digital solutions, especially those leveraging modern technologies like use of mobile apps for different service delivery could be reached. And true to their words, there are now numerous applications that offer real-time information on different issues, including traffic, weather, entertainment, government services, health, security and public utilities among other loads of information that one may be seeking (Angelidou et al. 2018). For instance, the Google Maps application, which can be accessed in different platforms including smartphones give real-time images of any part of the globe, including traffic, and this has been very instrumental for motorists and road users. Using the data from Google maps, other service providers are able to customize their applications also, such that they include information like location, distance and other pertinent issues relating to geographies of their businesses. Other examples that allow for real-time information include the e-citizen platform in Kenya that allow citizen to access government information and services in real-time from different devices, including mobile phones. Through this, they can pay for land rates, renew driving licenses and obtain immigration services just to name a few. *Waze* is another platform that has been very useful in giving real-time information on traffic updates to road users, and is hailed for allowing drivers to make prior planning on their travel routes by sharing information with other drivers (Waze 2019).

The possibilities that the P2P concept has especially with open source knowledge are immense and can bring unmetered transformation in urban areas. However, it may still be too early to predict how everything would pan out as for the moment, the smart cities and IoT concepts still rely on profitable direct business models, and ICT Corporations economically benefiting from the technology would object from opening up data as that would jeopardize their profits. Nevertheless, one cannot be wrong to propose and advocate for the open source knowledge sharing since this would unlock the innovation and creation in cities, and may even encourage the speeding up of implementation of the smart cities concept even in developing and less developed economies. Similarly, it would help cities offload the unemployment tag while also increasing their economic growth. Even more, with data available to all, novel approaches to address climate change and its challenges could be discovered as well as new and innovative ways of achieving the SDGs.

## REFERENCES

Alexander, C. (1965). A City is Not a Tree. *Architectural Forum, 122*(1), 58–61.
Alexander, C. (1979). *The Timeless Way of Building*. New York: Oxford University press.
Alexander, C. (2002). *The Nature of Order: The Process of Creating Life*. Berkeley, California: The Centre for Environmental Structure.

Alexander, C., Ishikawa, S., & Silverstein, M. (1977). *A Pattern Language*. New York: Oxford University Press.

Allam, Z. (2018). On Smart Contracts and Organisational Performance: A Review of Smart Contracts through the Blockchain Technology. *Review of Economic and Business Studies, 11*(2), 137–156.

Allam, Z. (2019a). Achieving Neuroplasticity in Artificial Neural Networks through Smart Cities. *Smart Cities, 2*(2), 118–134.

Allam, Z. (2019b). The Emergence of Anti-Privacy and Control at the Nexus between the Concepts of Safe City and Smart City. *Smart Cities, 2*(1), 96–105.

Allam, Z. (2020a). *Cities and the Digital Revolution: Aligning Technology and Humanity*. Springer International Publishing.

Allam, Z. (2020b). Data as the New Driving Gears of Urbanization. In Z. Allam (Ed.), *Cities and the Digital Revolution: Aligning Technology and Humanity* (pp. 1–29). Cham: Springer International Publishing.

Allam, Z. (2020c). Digital Urban Networks and Social Media. In Z. Allam (Ed.), *Cities and the Digital Revolution: Aligning Technology and Humanity* (pp. 61–83). Cham: Springer International Publishing.

Allam, Z. (2020d). Theology, Sustainability and Big Data. In Z. Allam (Ed.), *Theology and Urban Sustainability* (pp. 53–67). Cham: Springer International Publishing.

Allam, Z., & Dhunny, Z. A. (2019). On Big Data, Artificial Intelligence and Smart Cities. *Cities, 89*, 80–91.

Allam, Z., & Jones, D. S. (2019a). Attracting Investment by Introducing the City as a Special Economic Zone: A Perspective from Mauritius. *Urban Research & Practice, 12*, 1–7.

Allam, Z., & Jones, D. S. (2019b). The Potential of Blockchain within Air Rights Development as a Prevention Measure against Urban Sprawl. *Urban Science, 3*(1), 38.

Allam, Z., & Newman, P. (2018). Economically Incentivising Smart Urban Regeneration. Case Study of Port Louis, Mauritius. *Smart Cities, 1*(1), 53–74.

Allam, Z., Tegally, H., & Thondoo, M. (2019). Redefining the Use of Big Data in Urban Health for Increased Liveability in Smart Cities. *Smart Cities, 2*(2), 259–268.

Amoretti, M., & Zanichelli, F. (2018). P2P-PL: A Pattern Language to Design Efficient and Robust Peer-to-peer Systems. *Peer-to-Peer Networking and Applications, 11*(3), 518–547.

Angelidou, M., Psaltoglou, A., Komninos, N., Kakderi, C., Tsarchopoulos, P., & Panori, A. (2018). Enhancing Sustainable Urban Development through Smart City Applications. *Journal of Science and Technology Policy Management, 9*(2), 146–169.

Arias, R., Leuth, K. L., & Rastogi, A. (2018, January 21). The Effect of the Internet of Things on Sustainability. Retrieved from https://www.weforum.org/agenda/2018/01/effect-technology-sustainability-sdgs-internet-things-iot/

Bačić, Ž., Jogun, T., & Majić, I. (2018). Integrated Sensor Systems for Smart Cities. *Tehnički vjesnik, 25*(1), 277–284.

Barbosa, L., Pham, K., Silva, C., Vieira, M. R., & Freire, J. (2014). Structured Open Urban Data: Understanding the Landscape. *Mary Ann Liebert Inc Publishers [Online], 2*(3), 144–154.

Bassoo, V., Ramnarain-Seetohul, V., Hurbungs, V., Fowdur, T. P., & Beeharry, Y. (2018). Big Data Analytics for Smart Cities. In N. Dey, A. Hassanien, C. Bhatt, & S. Stapathy (Eds.), *Internet of Things and Big Data Analytics Toward Next-Generation Intelligence. Studies in Big Data* (Vol. 30). Cham: Springer.

Berntzen, L., Johannessen, M. R., & Florea, A. (2016). Smart Cities: Challenges and a Sensor-based Solution. *International Journal on Advances in Intelligence Systems, 9*(3&4), 579–588.

Bini, S. A. (2018). Artificial Intelligence, Machine Learning, Deep Learning, and Cognitive Computing: What Do these Terms Mean and How will they Impact Health Care? *The Journal of Arthroplasty, 33*(8), 2358–2361.

Blanco, J. L., Fuchs, S., Parsons, M., & Ribeirinho, M. J. (2018). Artificial Intelligence: Construction Technology's Next Frontier. Retrieved from McKinsey Website: https://www.mckinsey.com/industries/capital-projects-and-infrastructure/our-insights/artificial-intelligence-construction-technologys-next-frontier

Caird, S. P., & Hallett, S. H. (2018). Towards Evaluation Design for Smart City Development. *Journal of Urban Design, 24*, 1–22.

Danigelis, A. (2017, September 28). Smart City Market Growth Driven by Companies Like Schneider Electric and Ingersoll Rand. Retrieved from https://www.energymanagertoday.com/smart-city-market-growth-driven-companies-like-schneider-electric-ingersoll-rand-0171629/

Eden Strategy Institute, & ONG&ONG Pte. (2018). *Smart City Governments.* Retrieved from https://static1.squarespace.com/static/5b3c517fec4eb767a04e73ff/t/5b513c57aa4a99f62d168e60/1532050650562/Eden-OXD_Top+50+Smart+City+Governments.pdf

Ericsson. (2019). Internet of Things Forecast. Retrieved from https://www.ericsson.com/en/mobility-report/internet-of-things-forecast

Farhangi, H. (2010). The Path of the Smart Grid. *IEEE Power and Energy Magazine, 8*(1), 18–28.

Gellings, C. W., Samotyj, M., & Howe, B. (2004). The Future's Smart Delivery System [Electric Power Supply]. *IEEE Power and Energy Magazine, 2*(5), 40–48.

Hickins, M. (2017, December 12). Marseille Turns to Data to Plan a Safer City. Retrieved from https://www.forbes.com/sites/oracle/2017/12/12/marseille-turns-to-data-to-plan-a-safer-city/#1013f2121095

Huntingford, C., Jeffers, E. S., Bonsall, M. B., Christensen, H. M., Lees, T., & Yang, H. (2019). Machine Learning and Artificial Intelligence to Aid Climate Change Research and Preparedness. *Environmental Research Letters, 14*(12), 124007.

Ishida, T. (2017). *Digitial City, Smart City and Beyond.* Paper Presented at the 26th World Wide Web Intenational Conference (WWW17), Perth, Australia.

Jacobs, J. (1961). *The Death and Life of Great American Cities.* New York: Random House.

Joss, S., Sengers, F., Schraven, D., Caprotti, F., & Dayot, Y. (2019). The Smart City as Global Discourse: Storylines and Critical Junctures Across 27 Cities. *Journal of Urban Technology, 26*(1), 3–34.

Kamilaris, A., & Ostermann, F. O. (2018). Geospatial Analysis and the Internet of Things. *ISPRS International Journal of Geo-Information, 7*(7), 269.

Kamrowska-Zaluska, D., & Obracht-Prondzyńska, H. (2018). The Use of Big Data in Regenerative Planning. *Sustainability, 10*(10), 3668.

Keenan, M. (2018). The Future of Data With the Rise of the IoT. *RFID Journal,* 1–2.

Kim, T.-h., Ramos, C., & Mohammed, S. (2017). Smart City and IoT. *Future Generation Computer Systems, 76,* 159–162.

Liu, S. (2019, October 14). Global IoT Market Size 2017–2025. Retrieved from https://www.statista.com/statistics/976313/global-iot-market-size/

Lueth, K. L., & Pasqua, E. (2019). *IoT Solution World Congress 2019 Report.* Retrieved from Hamburg, Germany: https://iot-analytics.com/product/iot-solutions-world-congress-2019/

Manyika, J., Chui, M., Bisson, P., Woetzel, J., Dobbs, R., Bughin, J., & Aharon, D. (2015). *The Internet of Things: Mapping the Value Beyond the Hype.* Retrieved from https://www.mckinsey.com/business-functions/mckinsey-digital/our-insights/the-internet-of-things-the-value-of-digitizing-the-physical-world

Marzano, G., Lizut, J., & Siguencia, L. O. (2019). Crowdsourcing Solutions for Supporting Urban Mobility. *Procedia Computer Science, 149,* 542–547.

Mehaffy, M. W. The New Modernity: The Architecture of Complexity and The Technology of Life. Retrieved from http://www.katarxis3.com/Mehaffy_New_Modernity.htm

Mjimba, V., & Sibanda, G. (2019). Biomimicry, Big Data and Artificial Intelligence for a Dynamic Climate Change Management Policy Regime. In *Environmental Change and Sustainability.* Intech.

Qin, Y., Sheng, Q. Z., Falkner, N. J., Dustdar, S., Wang, H., & Vasilakos, A. V. (2016). When Things Matter: A Survey on Data-centric Internet of Things. *Journal of Networks and Computer Application, 64,* 137–153.

RideGuru. (2019). Rideshares Worldwide. Retrieved from https://ride.guru/content/resources/rideshares-worldwide#OLA

Rishi, R., & Saluja, R. (2019). *Future of IoT*. Retrieved from India https://www.ey.com/Publication/vwLUAssets/EY_-_Future_of_IoT/$FILE/EY-future-of-lot.pdf

Salingaros, N. (2010). Beyond Left and Right: Peer-to-Peer Themes and Urban Priorities for the Self-Organizing Society. Retrieved from https://patterns.architexturez.net/doc/az-cf-172685

Salingaros, N. (2011). *The Networked City and Peer-to-Peer Urbanism*. Paper Presented at the Talk given to the Institute for Advanced Architecture of Catalonia, Barcelona (IAAC): Iaac bits, Implementing Advanced Knowledge, Catalonia, Barcelona. Talk Retrieved from https://wiki.p2pfoundation.net/Brief_History_of_P2P-Urbanism

Salingaros, N., & Quintero, F. M. (2010, October). Brief History of P2P Urbanism. Retrieved from https://wiki.p2pfoundation.net/Brief_History_of_P2P-Urbanism

Salingaros, N. A. (1998). Theory of the Urban Web. *Journal of Urban Design, 3*, 53–71.

Salingaros, N. A. (2000). Complexity and Urban Coherence. *Journal of Urban Design, 5*, 291–316.

Salingaros, N. A. (2003). *Connecting the Fractal City*. Paper Presented at the 5th Biennial of towns and town planners in Europe, Barcelona.

Salingaros, N. A. (2006). Compact City Replaces Sprawl. In A. Graafland & L. Kavanaugh (Eds.), *Crossover: Architecture, Urbanism, Technology* (pp. 100–115). Rotterdam, Holland: 010 Publishers.

Salingaros, N. A. (2014). Complexity in Architecture and Design. *Oz Journal, 36*, 18–25.

Sarker, M. N. I., Wu, M., & Hossin, M. A. (2018, May 26–28). *Smart Governance Through Bigdata: Digital Transformation of Public Agencies*. Paper Presented at the 2018 International Conference on Artificial Intelligence and Big Data (ICAIBD), Chengdu, China.

Sasaki, M. (2010). Urban Regeneration through Cultural Creativity and Social Inclusion: Rethinking Creative City Theory through a Japanese Case Study. *Cities, 27*, S3–S9.

Shahab, S., & Allam, Z. (2020). Reducing Transaction Costs of Tradable Permit Schemes using Blockchain Smart Contracts. *Growth and Change, 51*, 302–308. https://doi.org/10.1111/grow.12342.

Souza, A., Figueredo, M., Cacho, N., Araújo, D., & Prolo, C. A. (2016). Using Big Data and Real-Time Analytics to Support Smart City Initiatives. *IFAC-PapersOnline, 49*(30), 257–263.

Statista Research Department. (2016, November 27). Internet of Things—Number of Connected Devices Worldwide 2015–2025. Retrieved from https://www.statista.com/statistics/471264/iot-number-of-connected-devices-worldwide/

Team Little. (2019). Little. Retrieved from https://www.little.bz/ke/

Thakuriah, P., Tilahun, N., & Zellner, M. (2017). Big Data and Urban Informatics: Innovations and Challenges to Urban Planning and Knowledge Discovery. In *Seeing Cities through Big Data: Research, Methods and Applications in Urban Informatics* (pp. 11–48). Springer.

Tomor, Z., Meijer, A., Michels, A., & Geertman, S. (2019). Smart Governance For Sustainable Cities: Findings from a Systematic Literature Review. *Journal of Urban Technology, 26*(4), 3–27.

Townsend, A. M. (2013). *Smart Cities: Big Data, Civic Hackers, and the Quest for a New Utopia*. New York, NY: W. W. Norton & Company.

Waze. (2019). Waze. Retrieved from https://ride.guru/content/resources/rideshares-worldwide#OLA

Zang, S., & Ye, M. (2015). Human Resource Management in the Era of Big Data. *Journal of Human Resource and Sustainability Studies, 3,* 41–45.

Zhang, D., & Qiu, R. C. (2018). Research on Big Data Applications in Global Energy Interconnection. *Global Energy Interconnection, 1*(3), 352–357.

# CHAPTER 4

# On Global Capitalism and Autonomous Smart Cities: A View on the Economic Engines of Tomorrow

**Abstract** Global structures are geared towards sustaining trade relationships that help economies achieve economic prosperity. Cities, both through their geographies and through their increasing economic role, help in this, and can even be seen to emerge as global superpowers, where some urban economic trump the economic performance of entire nations. As the role of capitalism gains in strength and cities reinforce their agenda as being global economic leaders, there will be a natural tendency towards models aimed towards better performance and efficiency in urban spheres. On this, the role of technology -coupled with urban governance is hailed. Smart cities is seen to aid but is limited in the sense that data is computed but await decisions from people—a step which can be cost deficit. This chapter supports that there will be a natural tendency to move towards autonomous cities to better support economic motives.

**Keywords** Autonomous cities • Capitalism • Urban economics • Future cities • Smart cities • Automation

© The Author(s) 2021
Z. Allam, *The Rise of Autonomous Smart Cities*, Sustainable Urban Futures, https://doi.org/10.1007/978-3-030-59448-0_4

# INTRODUCTION

Capitalism coincided with imperialism, where economic models prevailing in most of the economies, were used for selfish private interests, as vehicles to spur economic growth. This is vivid in the way colonialists actively extracted raw materials from their diverse colonies and exported the same to their countries where such were in high demand to feed industrialization (Allam 2019a; Allam and Jones 2019). Thereafter, after converting raw material into finished products, the surplus from the local market were sent to their colonies and exported to other markets with established trading partners. In the pursuit of those endeavors, capitalists, as noted in different quarters (Were 2018; Wagner 2018; Gilley 2017; Pwiti and Ndoro 1999; Morgan 2000; Austin 2010) fueled different kind of atrocities, pains and sufferings while enriching imperialists. The situation then were dire such that people like Karl Marx were concerned forcing them to agitate for changes, especially in respect to how the workers were treated by the owners of capital (Marx 1973; Marx and Engels 1885, 2018). In his view, capitalists impoverished workers by paying them meager wages and without improving their working conditions, while they themselves kept all the profits and benefits accrued.

In today's era of Post colonialism, surprisingly, capitalism is now being associated with democratic regimes, and is even being taunted as the currency of peace and stability. Gartzke (2007) explain that it is evident that democracies, especially those pursuing capitalist agendas seldom fight each other. Instead, they find ways to formulate policies that help each other's economic development, market integration and in pursuing common economic interests. On this, while the model is disputed, it is said to have had positive impacts on economic prosperity of different nations, rendering better liveability levels. Also, it has also been credited for increasing the pursuit of urbanization, and while this has its own shortcomings, it also has some positive impacts in the improvement of the economic performance of countries. This assertion is tied to the argument that cities are observed as the economic engines that run most economies. In today's fast urbanization pace, the economies of some cities have been observed to be relatively larger than those of many countries. For instance, in 2017, the city of Seoul had a bigger economy ($903 billion) than that of Malaysia ($817 billion) (Florida 2017). The city of Paris had a GDP of $755 billion in 2017 which was slightly higher than that of any African country during the same period (Gollain et al. 2019). Going into the future, fifteen years from now (2035), Ghosh (2019) projects that the world's major cities like New York ($2.5T), Tokyo ($1.9T), London ($1.3T), Shenzhen ($0.9T), and Paris ($1.1T) among many others will have superior economies in terms of GDP, and population growth than most countries.

Such increase in both population and economic growth will lead to higher stakes in national policies as city managers will have to balance between service provision, sustainability, resiliency and liveability status to name a few factors (Allam 2020). In line to this, it is observed that the use of cutting edge technologies will have a prominent role in ensuring the above balance is achieved. For instance, currently, most cities are turning towards the concept of smart cities to streamline operations, more so with the objective of increasing liveability status, resiliency and promoting sustainability, while ensuring higher performance and efficiency. On this, Causone et al. (2017) opine that such are achieved as this concept emphasizes on a variety of issues, where in our current times of stress from climate change, the concept is praised for its potential for reducing emissions. It also allows for prudent waste management, and in extension, leading to cleaner cities. Within the smart city environment, Zhou et al. (2016) explain that the issues of security, safety, efficiency and optimization of resource consumption are also given maximum attention. When all the benefits of this concept are compounded, they are seen to influence the liveability status of the city, which in turn attract inflow of talents, innovations, and investments; thus, influencing the economic standing of cities. For instance, the city of Seoul, South Korea is now positioned to offer its residents 'smart city' experiences in the transport sector, waste management and energy consumption and ability for them to communicate in a smart way to the government. The communication is enabled by availability of 120 *Dasan Call Centers* that allow citizens to enquire and communicate any issue with the metropolitan government.

With the changing technological terrain where novel additions like 5G, Artificial Intelligence (AI) and biotechnologies are being explored, the concept of smart cities is expected to expand beyond its current limits, and its influence on rendering cities into more efficient and liveable fabric is expected to be pursued (Allam et al. 2019; Dabeedooal et al. 2019). However, despite its positive impacts in various urban fabric, it is still limited in the fact that governance decisions are still in the hands of human beings (Viale Pereira et al. 2017; Sánchez-Corcuera et al. 2019). This is critical as humans, with their limitation, interests and personal and perhaps biased ambitions can, in one way or the other, derail the operations and impact the service delivery in cities, despite the presence of advanced technologies. Similarly, with the increasing global competitions, conflicts, and selfish interests, it is not unusual for the urban technological landscape to be infiltrated, destroyed or sabotaged, thus, impact core urban operations. Such eminent and possible scenarios are not to be taken lightly as any mishap in the city's fabric could have ripple effects on others areas, as the concept of smart city rides on the interconnectedness of devices, systems and networks (Allam 2019b). On such, to ensure that human limitations and their influences are minimized, the idea of moving to automation of cities is expected to gain traction. While this may be an unpopular scenario due to its potential to render cities as mechanical environments, it may be seen as a better complement for the human governance decisions in respect to climate change while influence better economic gains. Norman (2018) showcases that such gains would come from areas like cost savings where issues like delay, wastage, traffic congestions and insecurity can be minimized. Gains may also come from the increased knowledge sharing on different issues like renewable energy, urban health, and climate change mitigation strategies.

## MOVING PROFITS FROM CORPORATIONS TO CITIES

The smart city industry is an extremely profitable one, as showcased in Chap. 1, and this is set to increase even more with the projected increase in the number of Internet of Things (IoT). As noted by Horwitz (2019), the devices will not only increase to 75.44 billion pieces by 2025 from the estimated previous number of 26.6 billion devices reported in 2019, but they will also be powered by superior technologies with potential to optimize energy consumption, increase storage capacity and enhance compatibility issues. The unrelenting efforts geared toward enhancing the performance of these devices are on one hand seen geared on ensuring that the smart city concept is actualized and made even better. On the other hand, such are seen to be instigated by the increasing competition waged by the large ICT Corporations as they wrestle to outwit each other in the control of the IoT market. On this, both propositions hold weight.

First, in regard to focusing on actualizing the smart city concept, it is expected that future smart cities will benefit from superior technologies, requiring advanced IoT devices. This is pointed by the amount of investment injected in Research and Development (R&D) to produce devices that are superior, have increased bandwidth, and with capacity to capture and store infinite amount of data (Salimijazi et al. 2019; Panda et al. 2018; Adesina et al. 2017). Such desires have prompted even exploration of biotechnology frontiers, like exploitation of DNA and proteins for data and energy storage (De Silva and Ganegoda 2016; Zhao et al. 2019). Also, it is now evident that through the concept of Artificial Intelligence (AI), there are spirited efforts to mimic the human brain through Artificial Neuron Network (ANNs) (Huang 2017; Allam 2019b). Though the concept, as advanced by Zorins and Grabusts (2015), is still in its infancy stage, its full exploitation is projected to hasten the idea of automating cities, by reducing the amount of human interventions in different urban operations.

In regard to ICT Corporations influencing the smart city market for their own economic gains, there are various researchers pointing to this. First, Karakitsiou et al. (2017) shares how most of the devices installed in cities are uniquely designed and manufactured, such that, the run on unique protocols and networks which are seldom compatible with those of its competitors. The reasons for such strategy are multipronged. On one end, with incompatibility issues, city managers are force to contract specific corporations not only to install their products, but also for the operation and maintenance of the same. This ties the corporations to cities for longer periods than would be expected, and in such, it could be safe to argue that this leads to a case of economic hostage by ICT Corporations. Therefore, while cities are seen to derive substantial benefits from the use of tech in various dimensions of their systems, they are not the primary economic beneficiaries. ICT Corporations are. To put this into perspective, whenever a smart city project is initiated, whether it succeeds or not, contracted corporations are in most cases assured of their financial return. And, in cases where such projects are successful, they earn even more as they extend their stay through the operation and maintenance stages of the products they install. During all this time, city managers, planners and financing stakeholders have to grapple with the idea cost-benefit analysis, Return on Investments (ROIs) and break-even points among other such financial issues. An example of this is the Songdo city, where the process of making the city 'smart' started in 2008 under the Incheon u-city Corporation, and was planned to be complete by 2015, but the completion date was moved to 2022. During this time, corporations like Cisco, providing smart devices; sensors, motion detectors and camera among others continue to earn from their investment, but the local government is yet to break even from this $35 billion investment.

Additionally, it is noted that only a few cities around the globe can effectively and comfortably manage to finance such smart city projects without the need for external financiers. Theoretically, projects of smart city magnitude ought to be financed via government public spending. But, in reality, this has not been forthcoming as most economies are crowded by numerous financial obligations, such that they are left with little or no financial capacity to undertake capital intensive projects without external financial support. In respect to cities that have such capacity, factors like increasing urbanization and its associated risks prompt them to partner with the private sector or seek the support of International Development Organizations (IDOs) for such endeavors (Fishman and Flynn 2018; Flynn et al. 2018).

While such supports are welcome, and have allowed high-impact infrastructural projects to be undertaken in cities like Seoul and Barcelona, the burden to repay such are left on the shoulders of city managers, and noting that most of such attract long-term loans, they are argued to take substantially long before cities can break-even. Of concern is that most smart city projects are public projects undertaken to better and improve the urban fabric with no direct financial return expected. Such arguments justify that cities, despite having the projects installed within their perimeters cannot be argued to be the primary beneficiaries. But, while that is true, the indirect benefits that could be attributed to having smart cities as argued above are immeasurable. More importantly, their impacts in improving the liveability status and attractiveness of cities have far reaching positive impacts in influencing innovations, attracting Foreign Direct Investments (FDIs), attracting different forms of tourism and providing economic opportunities for urban residents.

While indirect benefits from smart city projects could suffice, there are possibilities of these cities becoming the ultimate economic beneficiaries right from the start as will be developed in the next section. The benefits for that are insurmountable and worth investing for. For instance, by being the ultimate beneficiary, it means that the projects are by a larger margin owned, run, maintained and monitored by urban managers. This would mean a reduction in the extra costs that are incurred to facilitate the contracted ICT Corporations. It also means that local companies and start-ups within the economy would secure opportunities to showcase their potential and that would have a huge bearing in job creation and creation of local solutions. Having control of the projects would also mean that the companies would have means and capacity to gather, store, process, analyze and share data without involving third party. The ability to safeguard the sanctity of data legitimizes the projects and in extension wins investor confidence, together with that of citizens and that would be a plus in boosting the economy of the city as expressed by Martinez-Balleste et al. (2013). This is further elaborated in the next section.

## AUTONOMOUS SMART CITIES
## AND ECONOMIC DEMOCRATIZATION

The role played by the private sector in actualizing infrastructure investments in any given city is unmatched, especially in respect to financial support, job creation, offering experts' technical support and promoting economic development. However, from an economic standpoint, it is true that profits made during the different phases of implementing such projects, including smart cities are enjoyed only by the few; more so those who are directly involved in said implementation. Here, the general public and the public administrators receive no direct benefits; and have to wait for the projects to be operationalized for them to start witness return on the investment. Such approaches to the implementation of public projects; though beneficial can be said to support segregated capitalistic motives where only a small group benefits from a market that is rapidly being monopolized by entrants with means and intentions to engulf demand. As noted by Ehlers (2014), large corporations have a fixation on winning as much control of this lucrative market as possible; hence, through their capacity to engage in extensive R&D, they flood the markets with products that small local companies and new startups cannot match. This strategy guarantees profits and market control. A case in point is Uber, whose dominance stifles efforts by local startups to establish themselves. It does this by using strategies like reducing base fares, a strategy that its competitors would not manage to integrate as they do not enjoy as much economies of scale as Uber does.

The desires by these large entrants to control as much market as possible breeds other economically unhealthy traits. First, it is understood that such barely allow for knowledge transfer, such that, by the time they complete the implementation of the projects, no local company can successfully take over the implementation and maintenance responsibility. This is justified by the way they code and package their products in a way that such are not compatible with other conventional products that have the capacity to use a universal network; hence allowing them to connect and communicate with other products of their likes. That strategy alone has the capacity to keep these large corporations in a city; managing and controlling a project that the local companies are technically apt to handle. By failing to transfer knowledge, this means that the local companies and startups will have problems accessing public data, which in such scenarios is still controlled by the large corporations. On this Tarnoff (2018) argues that when access to data is entrusted to a few profit motivated corporations, other stakeholders are left out of decision making, processes, even though if those may touch on their welfare. This scenario builds an unsustainable reliance on these large corporations.

The presence of large corporations' competition wars on small scale and startup local companies in an economy also means an unconducive environment to build capacity to match the increasing digital demands of cities. For instance, Tolica et al. (2015) observe that most of those companies cannot manage to enjoy economies of scale as they are usually financially constrained and find it hard to expand their scale of operations. That is tied to the difficulties the companies experience in attracting new talents that could inject new ideas and take forward the already existing technologically inclined concepts and ideas. Similarly, smaller scaled businesses do not have the capacity to invest in extensive R&D that would allow them to deliver superior products at par with those of their 'capitalistic' competitors. On this, Koutroumpis et al. (2019) highlight that small businesses and startups cannot afford to overlook the role of R&D as it dictates revenue scales and also the quality of products or services presented to the market. While those challenges are real, Koutroumpis et al. (2019) highlight that if such smaller firms get opportunities to grow and showcase their potentials, they have greater opportunities to dominate market niches, as they are more flexible and adaptable and are able to provide customized local solutions. However, the reality in most cities is that large corporations spearhead smart city designs both in developmental and operational stages. In such cases, this is seen to be leading to ultimate collapse of the local smaller companies and ultimately, affecting the local fabric negatively.

Without the input of local companies, it means that the large corporations can enjoy monopolistic control of the market, and this have far reaching impact in the influence, control, transparency and ethos of the industry. Unfortunately, there are fears that the implementation of the smart city concept is heading to this monopolistic path, where only large profit oriented corporations are involved in the large economically profitable ventures. An example here is the case of Uber ridesharing platform that has been accused in different parts of the globe of engaging in unfair competition, especially by influencing the pricing of the taxi industry; thus, exposing its main rivals in the industry to a myriad of challenges (Martínez 2019). The operation of Uber has been equated to the traditional form of capitalism, where the workers (drivers) earn meager returns while it keeps the remaining amount, though it owns no vehicle and do not bear the extra costs of maintaining the vehicles (Peticca-Harris et al. 2018; Fontana et al. 2019). This example is just a pointer of how smart digital solutions driven by large corporations have the potential to stifle local competitors and render the smart city concept as being part of monopolistic and capitalist enterprises.

To counter the above challenges, the rise of autonomous smart cities is projected to help as decisions regarding urban governance can be geared towards benefiting the whole fabric rather than solely private enterprises. The urban fabric will benefit as most of the decisions will not be based on capital driven interventions, but on the interactions between different smart devices with that of urban dwellers, fashioned to optimize the available resources for the benefit of urban landscape. For instance, with autonomous vehicles, Bouton et al. (2017) argue that smart cars will anticipate the commuter's need, and since they are in communication with other similar cars, will make the most optimal decision especially in regard to routing. A combination of such autonomous services in a city will lead to higher liveability status, increase efficiency and better economic performance. In particular, such will be even more effective if urban policies are formulated to provide enhanced control to the public administration rather than leaving them in the hands of private corporations. Doing this may help pave the way to a democratization of the economic landscape and in extension encourage entrepreneurship and a more socially inclusive environment as small businesses and startups will not fear being edged out of markets.

## References

Adesina, O., Anzai, I. A., Avalos, J. L., & Barstow, B. (2017). Embracing Biological Solutions to the Sustainable Energy Challenge. *Chem, 2*(1), 20–51.

Allam, Z. (2019a). *Theology and Urban Sustainability.* Springer International Publishing.

Allam, Z. (2019b). Achieving Neuroplasticity in Artificial Neural Networks through Smart Cities. *Smart Cities, 2*(2), 118–134.

Allam, Z. (2020). Digital Urban Networks and Social Media. In Z. Allam (Ed.), *Cities and the Digital Revolution: Aligning Technology and Humanity* (pp. 61–83). Cham: Springer International Publishing.

Allam, Z., & Jones, D. S. (2019). Climate Change and Economic Resilience through Urban and Cultural Heritage: The Case of Emerging Small Island Developing States Economies. *Economies, 7*(2), 62.

Allam, Z., Tegally, H., & Thondoo, M. (2019). Redefining the Use of Big Data in Urban Health for Increased Liveability in Smart Cities. *Smart Cities, 2*(2), 259–268.

Austin, G. (2010). African Economic Development and Colonial Legacies. *International Development Policy, 1*, 11–32.

Bouton, S., Hannon, E., Knupfer, S., & Ramkumar, S. (2017, June). The Future(s) of Mobility: How Cities Can Benefit. Retrieved from https://www.mckinsey.com/business-functions/sustainability/our-insights/the-futures-of-mobility-how-cities-can-benefit

Causone, F., Sangalli, A., Pagliano, L., & Carlucci, S. (2017). Assessing Energy Performance of Smart Cities. *Building Services Engineering Research and Technology, 39*(1), 99–116.

Dabeedooal, J. Y., Dindoyal, V., Allam, Z., & Jones, S. D. (2019). Smart Tourism as a Pillar for Sustainable Urban Development: An Alternate Smart City Strategy from Mauritius. *Smart Cities, 2*(2), 153–162.

De Silva, P. Y., & Ganegoda, G. U. (2016). New Trends of Digital Data Storage in DNA. *BioMed Research International, 2016*, 14.

Ehlers, T. (2014). *Understanding the Challenges for Infrastructure Finance*. BIS Working Papers (454, 454). Bank for International Settlements.

Fishman, T. D., & Flynn, M. (2018). Part Two: Funding and Financing Smart Cities Series. *Using public-Private Partnerships to Advance Smart Cities*, 1–10. Retrieved from https://www2.deloitte.com/content/dam/Deloitte/global/Documents/Public-Sector/gx-ps-public-private-partnerships-smart-cities-funding-finance.pdf

Florida, R. (2017, March 16). The Economic Power of Cities Compared to Nations. Retrieved from https://www.citylab.com/life/2017/03/the-economic-power-of-global-cities-compared-to-nations/519294/

Flynn, M., Rao, A. K., Horner, J., & Gashi, D. S. (2018). Smart Cities Funding and Financing in Developing Economies: Assisting Development Cites to Finance their Infrastructure Gap through Private Sector Participation Approaches. Retrieved from https://www2.deloitte.com/content/dam/Deloitte/global/Documents/Public-Sector/gx-ps-public-private-partnerships-smart-cities-funding-finance.pdf

Fontana, G., Pitelis, C., & Runde, J. (2019). Financialisation and the New Capitalism? *Cambridge Journal of Economics, 43*(4), 799–804.

Gartzke, E. (2007). The Capitalist Peace. *American Journal of Political Science, 51*(1), 166–191.

Ghosh, I. (2019, October 31). These Will be the Most Important Cities by 2035. Retrieved from https://www.weforum.org/agenda/2019/10/cities-in-2035/

Gilley, B. (2017). The Case for Colonialism. *Third World Quarterly*, 1–17.

Gollain, V., Tarquis, C., Burlin, Y., Saveli, I., Humbert, F., & Kutler, A. (2019). Retrieved from Mayenne, France: https://chooseparisregion.org/wp-content/uploads/2019/10/Paris-Region-Key-Figures-2019-BD.pdf

Horwitz, L. (2019, July 19). The Future of IoT Miniguide: The Burgeoning IoT Market Continues. Retrieved from https://www.cisco.com/c/en/us/solutions/internet-of-things/future-of-iot.html

Huang, T.-J. (2017). Imitating the Brain with Neurocomputer: A "New" Way Towards Artificial General Intelligence. *International Journal of Automation and Computing, 14*(5), 520–531.

Karakitsiou, A., Migdalas, A., Rassia, S. T., & Pardalos, P. M. (2017). *City Networks: Collaboration and Planning for Health and Sustainability*. Cham, Switzerland: Springer International Publishing.

Koutroumpis, P., Leiponen, A., & Thomas, L. D. W. (2019). Small is Big in ICT: The Impact of R&D on Productivity. *Telecommunications Policy*, 101833. doi.org/10.1016/j.telpol.2019.101833

Martínez, A. D. (2019). App Capitalism. *NACLA Report on the Americas, 51*(3), 236–241.

Martinez-Balleste, A., Perez-Martinez, P., & Solanas, A. (2013). The Pursuit of Citizens' Privacy: A privacy- Aware Smart City is Possible. *IEEE Communication Management, 51*(6), 136–141.

Marx, K. (1973). *Grundrisse*. Harmondsworth Penguine.

Marx, K., & Engels, F. (1885). *Das Kapital*. Hamburg: Verlag von Otto Meissner.

Marx, K., & Engels, F. (2018). *The Communist Manifesto*. Minneapolis, MN: Lerner Publishing Group, Inc..

Morgan, K. (2000). *Slavery, Atlantic Trade and the British Economy, 1660–1800*. Cambridge, UK: Cambridge University Press.

Norman, B. (2018). Are Autonomous Cities our Urban Future? *Nature Communications, 9*(1), 2111.

Panda, D., Molla, K. A., Baig, M. J., Swain, A., Behera, D., & Dash, M. (2018). DNA as a Digital Information Storage Device: Hope or Hype? *3 Biotech, 8*(5), 239.

Peticca-Harris, A., deGama, N., & Ravishankar, M. N. (2018). Postcapitalist Precarious Work and Those in the 'drivers' Seat: Exploring the Motivations and Lived Experiences of Uber Drivers in Canada. *Organization*. https://doi.org/10.1177/1350508418757332.

Pwiti, G., & Ndoro, W. (1999). The Legacy of Colonialism: Perceptions of the Cultural Heritage in Souther Africa, with Special Reference to Zimbabwe. *The African Archaeological Review, 16*(3), 143–153.

Salimijazi, F., Parra, E., & Barstow, B. (2019). Electrical Energy Storage with Engineered Biological Systems. *Journal of Biological Engineering, 13*(1), 38.

Sánchez-Corcuera, R., Nuñez-Marcos, A., Sesma-Solance, J., Bilbao-Jayo, A., Mulero, R., Zulaika, U., ... Almeida, A. (2019). Smart Cities Survey: Technologies, Application Domains and Challenges for the Cities of the Future. *International Journal of Distributed Sensor Networks, 15*(6). https://doi.org/10.1177/1550147719853984

Tarnoff, B. (2018, February 1). Data is the New Lifeblood of Capitalism—Don't Hand Corporate American Control. *Big Data*. Retrieved from https://www.theguardian.com/technology/2018/jan/31/data-laws-corporate-america-capitalism

Tolica, E. K., Sevrani, K., & Gorica, K. (2015). *Information Society Development through ICT Market Strategies: Albania versus Other Developing Countries.* New York, NY: Springer International Publishing.

Viale Pereira, G., Cunha, M. A., Lampoltshammer, T. J., Parycek, P., & Testa, M. G. (2017). Increasing Collaboration and Participation in Smart City Governance: A Cross-case Analysis of Smart City Initiatives. *Information Technology for Development, 23*(3), 526–553.

Wagner, K. A. (2018). Savage Warfare: Violence and Rule of Colonial Difference in Early British Counterinsurgency. *History Workshop Journal, 85*(Spring), 217–237.

Were, A. (2018). Debt Trap? Chinese Loans and Africa's Development Options. *South African Institute of International Affairs Policy Insights, 66*, 1–13.

Zhao, J., Wen, X., Xu, H., Wen, Y., Lu, H., & Meng, X. (2019). Salting-out and Salting-in of Protein: A Novel Approach toward Fabrication of Hierarchical Porous Carbon for Energy Storage Application. *Journal of Alloys and Compounds, 788*, 397–406.

Zhou, K., Fu, C., & Yang, S. (2016). Big Data Driven Smart Energy Management: From Big Data to Big Insights. *Renewable and Sustainable Energy Reviews, 56*, 215–225.

Zorins, A., & Grabusts, P. (2015). *Artificial Neural Networks and Human Brain: Survey of Improvement Possibilities of Learning.* Paper Presented at the 10th International Scientific and Practical Conference, Rezekne, Latvia.

# CHAPTER 5

# The Case for Autonomous Smart Cities in the Wake of Climate Change

**Abstract** Rapid urbanization coupled with climate change is causing a number of issues relating to urban governance and financing, which ultimately impacts on the liveability levels of urban areas. As the world turns towards smart cities to better increase economic performance of cities, there seems to be a lack of urban governance policy regarding climate change. Technology can aid on this front but must not be designed as a product solely for private economic gains, and instead designed to aid the public good. Smart cities, being a concept already making use of technology, can be redesigned to better make use of its technological background to ensure a better response to contemporary urban challenges, including that of climate change.

**Keywords** Autonomous cities • Smart cities • Sustainability • Climate change • Technology • Resilience • Urban policy • Future cities

## Introduction

Cities around the world occupies only 3% of the earth's surface area but are responsible for more than 75% of global carbon emissions (UN Environment Programme 2020; Satterthwaite et al. 2010). Despite occupying such a small area, they are home to over 54% of the global population, with the numbers projected to increase to 68–70% of 8.5 billion

© The Author(s) 2021
Z. Allam, *The Rise of Autonomous Smart Cities*, Sustainable Urban
Futures, https://doi.org/10.1007/978-3-030-59448-0_5

people that will be living on earth by 2030 (United Nations 2018). Such numbers are due to the positive reputation that cities have in respect to economic opportunities, liveability status and conglomeration of diverse quality services like education, health and transportation to name a few. They are also seen as safe havens relative to rural areas that are confronted by numerous challenges like reduced agricultural productivity following unpredictable weather conditions and out-migration of youthful population. Rural areas also offer limited opportunities in the economic and social spheres- in education, health and recreation. However, such are far much advanced in cities just as they are more accessible.

In respect to the above synopsis, it is worth noting that the liveability levels and opportunities being pursued in cities are unfortunately negatively impacted by the same economic drivers that make them attractive. That is, in the pursuit for economic growth, increased housing, construction of transport infrastructures and increase energy production to meet the surging demand among other related activities leads to massive emissions, excessive waste generation and excessive consumption of resources. Such negatives increase as more industries are constructed to increase production, more automobiles are introduced, rendering people more dependent on them, and demand for manufactured products fueled by changing consumption behaviors increases. Such trends are noted in the New Urban Agenda (NUA) (United Nations 2016) and others (UN Habitat 2011; IEA and UNEP 2018; UN Habitat 2018) as some of the urban challenges that cities have to accommodate with until there is a paradigm shift in the way issues pertaining to social inclusion, sustainability, resiliency and environmental protection are given due attention.

The need for the said paradigm shift is pegged among other issues; on the increasing role of cities in contributing to climate change, which in turn has contributed to numerous global challenges of the twenty-first century (Allam 2019). On this front, there are numerous regional and international agendas that are being pursued and advanced to address climate change and its impacts on the global sphere. Among those include the NUA, the SDGs especially number 11 (UN Environment Programme 2020), the Paris Agreement (United Nations 2015b), The Addis Ababa Action Agenda of the Third International Conference on Financing for Development (United Nations 2015a) and others promoted by United National Framework Convention on Climate Change (UNFCCC). The commonalities in these agendas are that they propose rafts of approaches, measures and interventions that have the potential to address matters relating to climate change, urban resilience and liveability status.

In line with the above agendas, an emerging concept that is anticipated to have potent solution in addressing climate change is that of smart cities, which is now being pursued in different cities across the globe and have shown flashes of promising outcomes in addressing the urban challenges including those related to climate change (Dabeedooal et al. 2019). Despite this, as noted by Tompson (2017), the concept is still in its infancy stages, and for this reason, only the private sector has had substantial confidence to finance and implement them on behalf of cities. For instance, a majority of the financing, supply of smart devices, systems, networks, and expertise have all been done by the private sector. While this is one way of bringing the concept into fruition, it has been widely criticized as being driven by profit agendas by said private enterprises. Nevertheless, for maximum benefits to be accrued by both the society at large and by urban administrators, the concept need to graduate from being a private enterprise to a social one in order to be embraced by all, and subsequently offer equal opportunities for all in the regions implemented.

As the challenges of climate change are accentuated, there is a need to revisit the smart cities concept. In particular, as will be argued in this chapter, the revisit is paramount as some aspects, especially certain urban governance decisions will need to be automated to allow for realistic opportunities to achieve better sustainability levels. This is possible as there are now advanced technologies based on Artificial Intelligence (AI) (McGovern et al. 2017; Anttiroiko et al. 2014) and Internet of Things (IoT) (Angelidou et al. 2018) that allow smart components to communicate via networks with minimum intervention of humans. Such have been demonstrated in areas like the transportation sector where spirited efforts and focus have been placed in producing autonomous vehicles that, among other things will not require human intervention. With such, as argued by Kim et al. (2012), they will help optimize resource consumption, as most of them, for instance, will be powered by renewable energy. They are also hailed for their potential to reduce air pollution (Rafael et al. 2020) and also reduce human related errors that usually leads to accidents, traffic congestion and many other issues (Kaur and Rampersad 2018). Full automation in this sector and others will see cities accrue even greater benefits like reduced costs, and reduce their over-reliance on profit-oriented private enterprises. The case of autonomous cars is one of many that demonstrates how automation of urban process can reduce human intervention and consequently, benefit urban resident for a higher urban liveability. As will be demonstrated in the succeeding section, automation can have greater impacts in even addressing issues of climate change on urban infrastructure.

## CLIMATE CHANGE AND URBAN INFRASTRUCTURAL LOSSES

The impacts of climate change are now more pronounced across the globe, as they are being experienced even in developed economies, though they have advanced infrastructure investments and being financially endowed to be able to implement mitigation programs. Issues like extreme temperatures (increased temperatures and extreme cold/hot weather) are now rampant in areas like Europe, Asia and even Africa without respect to developmental status. The temperatures have resulted into destruction of infrastructures, increased morbidity and in the worse-case scenarios, have resulted in losses of life. Such have prompted alarms from various quarters, including global organizations like UN Habitat (UN Habitat 2015), UNFCCC (Nations 2019) and other bodies (IEA and UNEP 2018; UNEP 2016; IPCC 2018). Subsequently, these have prompted the formulation of policies that target activities and processes in cities, and in entire economies at large. Among the most prominent ones include the New Urban Agenda (NUA), the Sustainable Development Goal (SDG) 11, the Sendai framework and Paris Agreement amongst many others. These in their own rights clearly denotes sustainability dimensions in line with addressing urban development to better respond to climate change and influence the liveability status in cities. These, especially the Sendai framework, also target the reduction of losses emanating from climate change disasters that are on the rise (UNDRR 2015). After the formulation of these policies, positive strides have been made across the globe and many countries are seen to increase their commitments toward reducing emissions that are responsible for climate change (UNDP 2010; Yeo 2019). For instance, in respect to the Paris Agreement, until today, 187 out of the 197 parties to the convention have ratified the agreement (UNFCCC 2016). In regard to the SDGs and NUA, it has been observed that countries have already amended some of their policies to align with these global policies (Sinha et al. 2020).

However, while the achievement in cities as a result of the above policy interventions are evident, they have been criticized as not being enough in view of increasing dramatic natural disasters (Simon et al. 2015). And, if the number of disasters, and their subsequent impacts in urban areas are to be respected, it would be safe to argue that the criticisms are justified. For instance, in 2019, almost five years after the Paris Agreement, some South African countries were hit by one of the harshest climate change instigated cyclones (Idai and Kenneth). Cyclone Idai alone left a trail of damages surpassing $773 million in infrastructure and more than 100,000 homes destroyed (World Vision 2019). Cyclone Kenneth was responsible for destruction of infrastructure worth more than $100 million in the same regions (AFDB 2019). Together, these are accused for having displaced over 2.2 million people and leaving more than 1,000 dead (UNCHA 2019). More recently, in Australia, in the wake of 2020, extreme heat and strong winds are being accused of dramatically escalating the wildfire disaster that have engulfed the continent claiming over 24.5 million acres of land (BBC 2020), killing over a billion animals (Samuel 2020), and at the time of writing this chapter, more than 28 people were reported to have lost their lives (BBC 2020). A report by Burbank et al. (2014) showcased that due to extreme heats, over 70,000 lost their lives in Europe in 2003. In Vermont, US, the report highlighted that in 2011, a tropical storm (Irene) led to the destruction of over 2,000 roads, 1,000 culverts and approximately 200 miles of railway and over 200 bridges to be closed resulting into major transportation problem. The above figures represent only a very small percentage of what the world is experiencing as events instigated by climate change keep on increasing; thus, justifying the calls for increased urgency in addressing matters climate change.

It is noted that cities located within coastal regions are particularly more vulnerable to climate change instigated events due to their location. On this, it has been found that these areas experience frequent storms, strong winds, high temperatures and flooding relative to their counterpart located inlands (UN Habitat 2015; Cottrell et al. 2015). Such happenings are mostly influenced by human induced activities like the over-reliance on non-renewable energy sources, over-exploitation of coastal resources and other activities that causes erosions and many others that in turn have led to the current challenge of climate change.

While the above issues paint a troubling situation, those impacts are not seen to halt in the foreseeable future unless urgent and practical interventions are sought, especially with an aim of preparing urban communities to better respond to climate change. In particular, one of the interventions being pursued is the investment in 'infrastructural resilience', where those areas -especially coastal regions, are encouraged to install infrastructural programs that could help them withstand the wrath of climate change events. This call have been heeded, and as World Bank (2010) reports, there are spirited efforts even in less developed economies to invest in infrastructural development despite financial constraints. On this line, noting that most infrastructural investments are capital intensive; hence, prompting many economies to plunge into debts; deploying novel technologies like smart cities could help in getting the best out of those projects (Chironga et al. 2018; Kuwonu 2016; Lee 2014; UN-Habitat 2014) as will be discussed in the section below. Such technologies, despite being expensive may guarantee valuable assets and may also spur economic development and growth that could, in the long-run, help in meeting debt repayment obligations. Additionally, those have the potential to address notable climate change impacts, especially through early detection and prediction that are made possible by AI technologies like Machine Learning.

## CLIMATE CHANGE AND URBAN TECHNOLOGIES

As the smart city market continue to enlarge, ICT Corporations are seen to also increase their activities in the field. In particular, they are seen to invest massively in R&D, such that the products and services they present in the smart cities are of distinct and of high quality; all geared toward winning competitive advantages over others. Globally, in 2019 it is reported that ICT firms invested approximately $239 billion in R&D; a substantial increase from $228.3 billion that was invested in 2018. In 2017, the investments in this sector amounted to $218.3 billion, which was approximately $10 billion increase from the 2016 recorded figure of $207.7 billion (Duffin 2019). These investments are partly influenced by the rate at which the smart city concept and that of urban digital solutions have been growing especially in its capitalization. In 2018, IMARC (2020) Group argues that the market was valued at US $312.4 billion and projected that it would reach US$826.3 billion by 2024 at an expected growth rate of 17.6% in the period 2019–2024. Another report by Persistence Market Research valued the global smart cities market at US$622 billion and estimate that it would grow to approximately US$3.48 billion by 2026 (Smart Cities Association 2020). Even if those figures differ, more so due to differing methodologies, those show that the smart cities market are influenced by the global increase in government investment in urban digital solutions to remain abreast with the unprecedented rate of increase in urbanization.

In view of the above, there is an increase in investments in smart-city-oriented digital solutions focusing on areas like weather forecasting, biking, smart irrigation, smart waste management solutions and others geared toward rendering cities more resilient. These, in their own rights have been seen to bring outstanding changes on how the urban fabric is structured. For instance, in respect to weather forecasting, such technologies like the use of mobile apps to monitor weather have been further developed and are helping in urban decision making in areas like transport, water usage, and in mitigation strategies to name a few. Bauer et al. (2015) express that using available urban data, it is now possible to conduct numerical weather predictions, which are similar to brain simulations. McGovern et al. (2017) explain that this is possible due to the integration of prediction methods powered by AI technologies, which are helping cities make real-time decision on areas like investment in alternative energy. With such predictions, cities are seen to increase avenues for savings on their costs; which is paramount in increasing the efficiency of urban management. Further to this, with smart irrigation, Cano et al. (2018) express how it is now possible to remotely control irrigations systems; thus, improve in novel water saving technique, ensure public green spaces are well managed and soil moistures are monitored. Such technologies have been deployed in smart city Barcelona and is hailed for the way it has achieve remarkable outcomes in management of water networks and resources Libelium (2016).

While most urban smart technologies have proven to achieve unquestionable outcomes in making cities resilient and liveable, they still heavily rely on data. On this, though most are able to allow for easy collection and analysis of data and subsequently provide deep insights, they are in most cases unequipped with the power to generate actionable results. Therefore, such rely on human interventions, especially in respect to interpretation for actionable outputs. Here, the real issue of maximization of benefits of smart cities arises, particularly for relying on human decision making. From literature, most smart cities data are collected and handled by third parties, especially the profit-oriented enterprises. Those ICT companies (often international monopolies) cause friction with local companies and startups by denying them total access to data. In other cases, they have been accused of commercializing the data for their sole private gains. Compounding all these data challenges serve as pointers to challenges that may arise when dealing with increasing performance of resilience programs and efficiency of disaster responses. Among those challenges is the universal acceptability of the projects, especially by the citizens who may feel cheated if their data is to be controlled by third parties, of whom they may not have privy information of their mandate in smart city implementation. The local companies and startups may also be unable to offer solutions and this would expose the market to total exploitation by the private sectors and all those in control of said data.

Therefore, with that background, it would be necessary to ponder on the imperative of how to render more efficient resilience programs through automation in respect to climate change; where on this the democratization of technology is key in order to render a more inclusive economic landscape.

## On the Ethics of Climate Action and Urban Policies

In view of the increasing challenges on climate, and the calls by global organizations, local governments have the imperative to work on policy decisions and their enactment. This is further accentuated since the consequences of climate change are heavily endured in urban areas, resulting indirect impacts on the economy, social sphere, the environment, and political landscape. On the economic front, the vulnerability of urban infrastructures (Forzieri et al. 2018), the urban fabric and other aspects of urban areas are enough to cause a reversal or slowdown of economic achievements, rendering unprecedented losses and negative outcomes. For instance, with wavering economic environments characterized by insufficient, unreliable, and dilapidated infrastructures, it would be hard for a city to attract investors, new talents that spur innovations, or tourists. In addition, such infrastructures, as explained by Moretti and Loprencipe (2018), can attract extra costs like those of maintenance and construction, linked to those new infrastructural structures. In addition, such are seen to plunge cities into debt cycles as local managements are pressured to secure loans from different sources to address the infrastructural shortages. More still, urban areas are forced to incur costs associated with rebranding as they try to win back the confidence of investors, tourists and other stakeholders who may have shifted their attention, investments, and interests somewhere else. Here, climate change is understood to cause problems like the displacement of people, whom, in most cases are seen to end up in informal settlement areas. Others are forced in the outcasts of the cities;

thus, increasing the problems of urban sprawl. In extreme events, those have been seen to lead to loss of lives and injuries in different cities across the globe. The ripple effects of such is increased costs in health, negative impacts on the economy and on the reputation of the city.

In respect to the environment, it is evident that the current situation have been worsening with climate change consequences such as extreme temperatures, flooding, rise in sea levels, loss of biodiversity and emergence of new invasive species being witnessed. Politically, the issue of climate change has seen an escalation of disharmony between different countries, as a disagreement on issues related to environmental responsibility, decarbonization, reduction of emissions and commitment to international agreements continue to be observed (Zhang et al. 2017).

The fact that all those consequences of climate change on communities and impacts on their survival highlight an ethical imperative where speedy action is needed; supporting that urban policies need to be realigned in order to better respond to this global urgency. In this case the SDG 11 and the NUA provide good guidelines on what needs to be done to ensure the plight of locals are factored in when talking development. Such need to inform the local realignment, such that the policies are focused on ensuring that those two global policies are achieved. On this, the adoption and use of advanced smart urban technologies can help achieve the objectives set in the two policy documents. From the literature, it is clear that diverse technologies such as IoT, AI (Calo 2017), Big Data (Ivanov and Gnevanov 2018; Osman et al. 2017; Batty 2012), Crowd Computing and ANNs (Kotenko et al. 2015; Lee et al. 2016; Huang 2017) among many others are available to be harnessed not only to make urban areas smart, but also to help in automation of processes and activities. By so doing, it would be possible to minimize, or even eliminate human interventions in decision making that have been argued to distort the objectives set in both global and local climate change policies. The pursuit of automation of cities need to be hastened, especially if that can provide a lifeline in the actualization of the Paris Agreement, allow for quicker implementation of SDG 11 and of other global proposed interventions mentioned in this document.

While contemporary literature may portray the smart city concept emphasized in this chapter as a futuristic scenario, its promises in making urban areas more liveable, resilient, sustainable and socially inclusive. On this, the pertinent question that need to be asked is when and how the concept can be adopted and implemented fully in all cities so that the deep ethical and moral issues that appertains to communities, access and usage of data and the role of third parties in managing same can be addressed comprehensively. Macrorie et al. (2019) highlight that the automation of cities, especially through the use of robots and machines could help reduce the apprehensiveness that people have in sharing their data as in most cases, such are handled by the third party. Here, though Macrorie et al. (2019) acknowledges that automation also raise some concerns, it is possible to address such if there is openness and transparency in the handling of data collected from the urban fabrics and from the citizenry found in these cities.

As the benefits of technology in cities are apparent, and the challenges of climate change are highlighted and seen to impact on the livelihood of urban, rural and coastal communities, it can be argued that it is an ethical imperative to adopt the concept of automation of smart cities to ensure that more timeline and efficient solutions are adopted in response to complex climate issues.

## REFERENCES

AFDB. (2019). *Programme Post Cyclone Idai and Kenneth Emergency Recovery and Resilience Programme for Mozambique, Malawi and Zimbabwe (PCIREP).* Retrieved from https://www.afdb.org/fileadmin/uploads/afdb/Documents/Project-and-Operations/Multinational_-_Programme_post_cyclone_Idai_and_Kenneth_emergency_recovery_and_resilience_programme_for_Mozambique__Malawi_and_Zimbabwe_%E2%80%93__PCIREP__-_Appraisal_report.pdf

Allam, M. Z. (2018). *Redefining the Smart City: Culture, Metabolism and Governance. Case Study of Port Louis, Mauritius* (PhD), Curtin University, Perth, Australia. Retrieved from https://espace.curtin.edu.au/handle/20.500.11937/70707

Allam, M. Z. (2019). *Urban Resilience and Economic Equity in an Era of Global Climate Crisis.* University of Sydney.

Angelidou, M., Psaltoglou, A., Komninos, N., Kakderi, C., Tsarchopoulos, P., & Panori, A. (2018). Enhancing Sustainable Urban Development through Smart City Applications. *Journal of Science and Technology Policy Management, 9*(2), 146–169.

Anttiroiko, A. V., Valkama, P., & Bailey, S. J. (2014). Smart Cities in the New Service Economy: Building Platforms for Smart Services. *Artificial Intelligence and Society, 29*, 323–334.

Batty, M. (2012). Big Data, Smart Cities and City Planning. *Environment and Planning B: Planning and Design, 39*, 191–193.

Bauer, P., Thorpe, A., & Brunet, G. (2015). The Quiet Revolution of Numerical Weather Prediction. *Nature, 525*(7567), 47–55.

BBC. (2020, January 21). Australia Fires: A Visual Guide to the Bushfire Crisis. Retrieved from https://www.bbc.com/news/world-australia-50951043

Burbank, C. J., Kuby, M., Oster, C., Posey, J., Russo, E. J., Rypinski, A. ... Brinckerhoff, P. (2014). *National Climate Assessment-Transportation*. Retrieved from Washington, DC: https://nca2014.globalchange.gov/report/sectors/transportation

Calo, R. (2017). Artificial Intelligence Policy: A Roadmap. *SSRN Electronic Journal*, 1–28. https://doi.org/10.2139/ssrn.3015350

Cano, L., Ortega, C., Talavera, A., & Lazo, J. G. L. (2018). Smart City Park Irrigation System: A Case Study of San Isidro, Lima—Peru. *Proceedings, 2*(19), 1227.

Chironga, M., Cunha, L., De Grandis, H., & Kuyoro, M. (2018). *Roaring to Life: Growth and Innovation in African Retail Banking*. Retrieved from https://www.mckinsey.com/~/media/mckinsey/industries/financial%20services/our%20insights/african%20retail%20bankings%20next%20growth%20frontier/roaring-to-life-growth-and-innovation-in-african-retail-banking-web-final.ashx

Cottrell, J., Fortier, F., & Schlegelmilch, K. (2015). *Fossil Fuel to Renewable Energy: Comparator Study of Subsidy Reforms and Energy Transitions in African and Indian Ocean Island States*. Retrieved from Incheon, Republic of Korea: https://unosd.un.org/sites/unosd.un.org/files/ffre_islands_comparator_study_2015_printed_version_0.pdf

Dabeedooal, J. Y., Dindoyal, V., Allam, Z., & Jones, S. D. (2019). Smart Tourism as a Pillar for Sustainable Urban Development: An Alternate Smart City Strategy from Mauritius. *Smart Cities, 2*(2), 153–162.

Duffin, E. (2019, April 29). ICT Research and Development Expenditure in U.S. and Worldwide 2015–2018. Retrieved from https://www.statista.com/statistics/732308/worldwide-research-and-development-information-communication-technology/

Forzieri, G., Bianchi, A., Silva, F. B. e., Marin Herrera, M. A., Leblois, A., Lavalle, C., ... Feyen, L. (2018). Escalating Impacts of Climate Extremes on Critical Infrastructures in Europe. *Global Environmental Change, 48*, 97–107.

Huang, T.-J. (2017). Imitating the Brain with Neurocomputer: A "New" Way Towards Artificial General Intelligence. *International Journal of Automation and Computing, 14*(5), 520–531.

IEA and UNEP. (2018). *2018 Global Status Report: Towards a zero-emission, Efficient and Resilient Buildings and Construction Sector*. Retrieved from https://www.worldgbc.org/news-media/2018-global-status-report-towards-zero-emission-efficient-and-resilient-buildings-and

IMARC. (2020). Smart Cities Market: Global Industry Trends, Shae, Size, Growth, Opportunity and Forecast 2019–2024. Retrieved from https://www.imarcgroup.com/smart-city-market

IPCC. (2018). Summary for Policymakers. In V. Masson-Delmotte, P. Zhai, H. O. Pörtner, D. Roberts, J. Skea, P. R. Shukla, A. Pirani, W. Moufouma-Okia, C. Péan, R. Pidcock, S. Connors, J. B. R. Matthews, Y. Chen, X. Zhou, M. I. Gomis, E. Lonnoy, T. Maycock, M. Tignor, & T. Waterfield (Eds.), *Global Warming of 1.5°C. An IPCC Special Report on the Impacts of Global Warming of 1.5°C above Pre-industrial Levels and Related Global Greenhouse Gas Emission Pathways, in the Context of Strengthening the Global Response to the Threat of Climate Change, Sustainable Development, and Efforts to Eradicate Poverty.* Geneva, Switzerland: World Meteorological Organization.

Ivanov, N., & Gnevanov, M. (2018). Big Data: Perspectives of Using in Urban Planning and Management. *MATEC Web of Conferences, 170*(01107), 1–5.

Kaur, K., & Rampersad, G. (2018). Trust in Driverless Cars: Investigating Key Factors Influencing the Adoption of Driverless Cars. *Journal of Engineering and Technology Management, 48*, 87–96.

Kim, T. S., Na, J. C., & Kim, K. J. (2012). Optimization of an Autonomous Car Controller Using a Self-Adaptive Evolutionary Strategy. *International Journal of Advanced Robotic Systems, 9*(3), 73.

Kotenko, I., Saenko, I., Skorik, F., & Bushuev, S. (2015). *Neural Network Approach to Forecast the State of the Internet of Things Elements.* Paper Presented at the XVIII International Conference on Soft Computing and Measurements (SCM).

Kuwonu, F. (2016). *Africa's Cities of the Future.* Retrieved from New York, NY: https://www.un.org/africarenewal/magazine/april-2016/africa%E2%80%99s-cities-future

Lee, D. (2014). *Africa is Ready to Leapfrog the Competition Through Smart Cities Technologies.* Retrieved from https://www2.deloitte.com/content/dam/Deloitte/za/Documents/risk/ZA_SMARTCITIESA4(VIEW)_020615.pdf

Lee, K. Y., Chung, N., & Hwang, S. (2016). Application of an Artificial Neural Network (ANN) Model for Predicting Mosquito Abundances in Urban Areas. *Ecological Informatics, 36*, 172–180.

Libelium. (2016, August 29). Saving Water with Smart Irrigation System in Barcelona. Retrieved from http://www.libelium.com/saving-water-with-smart-irrigation-system-in-barcelona/

Macrorie, R., Marvin, S., & While, A. (2019). Robotics and Automation in the City: A Research Agenda. *Urban Geography*, 1–21. https://doi.org/10.1080/02723638.2019.1698868

McGovern, A., Elmore, K. L., Gagne, D. L., II, Haupt, S. E., Karstens, C. D., Lagerquist, R., ... Williams, J. K. (2017). Using Artificial Intelligence to

Improve Real-Time Decision-Making for High-Impact Weather. *Bulletin of the American Meteorological Society, 98*(10), 2073–2090.

Moretti, L., & Loprencipe, G. (2018). Climate Change and Transport Infrastructures: State of the Art. *Sustainability, 10*(11), 4098.

Nations, U. (2019). United Nations Framework Convention on Climate Change: Timeline. Retrieved from http://unfccc.int/timeline/

Osman, A. M. S., Elragal, A., & Bergvall-Kareborn, B. (2017). *Big Data Analyticas and Smart Cities: A Loose or Tight Couple?*, Lulea, Sweden: Luleå University of Technology.

Rafael, S., Correia, L. P., Lopes, D., Bandeira, J., Coelho, M. C., Andrade, M., ... Miranda, A. I. (2020). Autonomous Vehicles Opportunities for Cities Air Quality. *Science of The Total Environment, 712*, 136546.

Samuel, S. (2020, January 7). A Staggering 1 Billion Animals are Now Estimated Dead in Australia's Fires. Retrieved from https://www.bbc.com/news/world-australia-50951043

Satterthwaite, D., McGranahan, G., & Tacoli, C. (2010). Urbanisation and Its Implications for Food and Farming. *Philosophical Transactions of the Royal Society B: Biological Sciences, 365*(1554), 2809–2820.

Simon, D., Arfvidsson, H., Anand, G., Bazaz, A., Fenna, G., Foster, K., ... Wright, C. (2015). Developing and Testing the Urban Sustainable Development Goal's Targets and Indicators—a Five-city Study. *Environment and Urbanization, 28*(1), 49–63.

Sinha, A., Sengupta, T., & Alvarado, R. (2020). Interplay between Technological Innovation and Environmental Quality: Formulating the SDG Policies for next 11 Economies. *Journal of Cleaner Production, 242*, 118549.

Smart Cities Association. (2020). Global Smart Cities Market to Reach a Whopping $3.5 Trillion by 2026. Retrieved from https://www.smartcitiesassociation.org/index.php/media-corner/news/1-global-smart-cities-market-to-reach-a-whopping-3-5-trillion-by-2026

Tompson, T. (2017). Understanding the Contextual Development of Smart City Initiatives: A Pragmatist Methodology. *She Ji: The Journal of Design, Economics, and Innovation, 3*(3), 210–228.

UN Environment Programme. (2020). Goal 11: Sustainable Cities and Communities. Retrieved from https://www.unenvironment.org/explore-topics/sustainable-development-goals/why-do-sustainable-development-goals-matter/goal-11

UN Habitat. (2011). *The Economic Role of Cities*. Retrieved from Nairobi: http://urban-intergroup.eu/wp-content/files_mf/economicroleofcities_unhabitat11.pdf

UN Habitat. (2015). *Urbanization and Climate Change in Small Island Developing States*. Retrieved from Nairobi, Kenya: https://unhabitat.org/wpdm-package/urbanization-and-climate-change-in-small-island-developing-states/?wpdmdl=114762

UN Habitat. (2018, October 31). World Cities Day: Building Sustainable and Resilient Cities. Retrieved from https://unhabitat.org/wcd-2018/

UNCHA. (2019, December 10). Cyclones Idai and Kenneth. Retrieved from https://www.unocha.org/southern-and-eastern-africa-rosea/cyclones-idai-and-kenneth

UNDP. (2010). *UNDP Community Water Initiative: Fostering Water Security and Climate Change Adaptation and Mitigation.* New York, NY: United Nations Development Programme (UNDP).

UNDRR. (2015). *Sendai Framework for Disaster Risk Reduction* (UNDRR Ed.). Sendai City, Miyagi Prefecture, Japan: United Nations.

UNEP. (2016). *Global Material Flows and Resource Productivity.* An Assessment Study of the UNEP International Resource Panel (H. Schandl, M. Fischer-Kowalski, J. West, S. Giljum, M. Dittrich, & N. Eisenmenger, Eds.). Paris: United Nations Environment Program.

UNFCCC. (2016). Paris Agreement—Status of Ratification. Retrieved from https://unfccc.int/process/the-paris-agreement/status-of-ratification

UN-Habitat. (2014). *State of African Cities 2014, Re-imagining Sustainable Urban Transitions.* Retrieved from Nairobi: https://unhabitat.org/state-of-african-cities-2014-re-imagining-sustainable-urban-transitions

United Nations. (2015a). *Addis Ababa Action Agenda of the Third International Conference on Financing for Development (Addis Ababa Action Agenda).* Paper Presented at the Third International Conference on Financing for Development Addis Ababa, Ethiopia.

United Nations. (2015b). Paris Agreement. *UNFCCC.* Retrieved from UNFCCC website: https://unfccc.int/sites/default/files/english_paris_agreement.pdf

United Nations. (2016). *The New Urban Agenda.* Paper Presented at the United Nations Conference on Housing and Sustainable Urban Development (Habitat III), Quito, Ecuador.

United Nations. (2018). World Urbanization Prospects-The 2018 Revision. *Population Division.* Retrieved from https://population.un.org/wup/Publications/Files/WUP2018-Report.pdf

World Bank. (2010, December). *Cities and Climate Change: An Urgent Agenda.* Urban Development Series Knowledge Papers (63704, 10). Urban Development Series Knowledge Papers, Washington, DC.

World Vision. (2019). 2019 Cyclone Idai: Facts, FAQs, and How to Help. Retrieved from https://www.worldvision.org/disaster-relief-news-stories/2019-cyclone-idai-facts

Yeo, S. (2019). Where Climate Cash is Flowing and Why it's Not Enough. *Nature, 573*(7774), 328–331.

Zhang, Y.-X., Chao, Q.-C., Zheng, Q.-H., & Huang, L. (2017). The Withdrawal of the U.S. from the Paris Agreement and Its Impact on Global Climate Change Governance. *Advances in Climate Change Research, 8*(4), 213–219.

# CHAPTER 6

# Global Tourism and the Risks of Cultural Homogeneity in Smart and Future Cities

**Abstract** With technological innovation gaining ground, and their adoption in cities all around the world are being witnessed and documented, their impacts on the lifestyles of people are increasingly being looked at. Beyond the immediate fields in which those technologies operate, various aftermaths are being observed in indirect but related quarters. One of this is the cultural domain with the intersection of the global tourism industry—which is raising the demand for cultural products and locations, which on the end technological products associated smart cities are clashing against the richness of traditional urban areas. The rapid pace of technological adoption can lead to homogeneity in the urban realm and further erase the identity of places if this is not looked at. This final chapter explores this issue and offers perspectives on how the cultural dimension is important and cannot be made a collateral in the digital revolution that most cities are facing.

**Keywords** Cultural cities • Smart cities • Homogeneity • Autonomous cities • Culture • Future cities

Z. Allam, *The Rise of Autonomous Smart Cities*, Sustainable Urban Futures, https://doi.org/10.1007/978-3-030-59448-0_6

## INTRODUCTION

As from late twentieth and early twenty-first century to date, the world has been experiencing trends and phenomenon of momentous magnitudes, relating to the subsequent advancement of technology. Those have prompted positive disruptions in diverse sectors; in particular to urban areas, with a focus on themes such as, population increase and the new challenges of climate change amongst many others. These, on their own rights have impacted on sectors like health, education, business, transport and communication, energy and tourism, environment and local and global politics amongst others. For instance, looking at the tourism sector, it is evident that the number of those making use of global travels, with the greater accessibility to remote regions at better pricing has significantly increased. On this, the World Tourism Organisation (2012) highlights the global annual number of travelers in 1950 was 25 million, but by 2011, this figure had increased to more than 990 million, and its positive associated trends have been ongoing. This is partly seen to have been influenced by the advancement in technology in transportation industries; especially air transport, which has benefited significantly. Similarly, such travels have been accentuated by the exponential growth and increase in disposable income in both industrialized and emerging economies, accelerated by globalization. The said trends in this sector have affirmed that the potentials in the travel industry have direct impacts on urban areas. This finding is of high importance as travels are in most cases directed at cities and when that is not the case, the travelling population still heavily interacts with the urban fabric as most major transportation hubs are based in cities.

In view of this potential, it has been observed that most local governments and urban managers have increased their effort, attention and investments in ensuring they maximize the benefits that come along with tourism. In particular, there is evidence of increased attention in digital infrastructures that have the capacity to support the new influx of visitors (Allam 2020a, b, c). Such investments are customized and tailored toward maximizing the experience of visitors deliberately and geared to ensuring that as the visitors enjoy their stay in the city. In doing this, the economic situation therein is positively influenced and the return on such investments are increased. Here, Barbier et al. (2017) support that digital solutions like mobile applications (apps) for car sharing, especially in the taxi business have brought positive transformations both in the urban transport sector and in the automobile industry. For instance, these have been instrumental in cutting costs on parking, municipal taxi licenses and also in reducing traffic congestions. In regard to tourism and transient visits from the local population, those apps provide new levels of experience as most of them allow for customization such that one can tailor their settings to accommodate those with their own way of life and the language they are conversant with. Besides taxi services, mobile apps and websites allowing travelers to book and schedule their travel in any part of the world are increasing. This is a plus as those address time wastage in queues or during travel to agencies to acquire booking services. Even better, these technologies are also allowing visitors to pay for said services online or via mobile money transfer; thus, reducing incidences of cash payments; which are expensive, especially in cases of foreign currencies.

However, while there is a notable advancement in the travel industries—warranted by technological changes, some interesting studies (Böcker et al. 2017; Nakanishi and Black 2016) have shown that the convenience experienced here may be more popular with the elder generation. This is so as millennials have been observed to give more value to cultural experiences, entailing travelling for meaningful purposes, including personal development or personal explorations. For the elder generations, convenience and security is highly rated, and are seen to be more inclined to enjoy and like material goods. Millennials, on the other hand, have their travel decisions influenced and dictated by the unique cultural identities and experiences on offer in varying cities. To support this, Fromm (2018) denotes that this demographic group is seen to be attracted by memories and events; which, are even better in the modern days due to the availability of diverse social media platforms where such can be shared. In addition, the availability of mobile devices with the potential to capture and share such moments in real time to a wider audience provide millennials the impetus to enjoy said cultural experiences. The quest for new such experiences means that millennials are comfortable to travel locally; even in remote areas as long as they are able experience momentous and unique cultural attractions. To them, the locality of such things like attractive, art, nature, culinary experiences and other local materials suffice to feed their travel desires.

Such highlights by the youthful generation on travel decision need not be taken lightly by the management of cities, as those represent business ventures. However, they should take it as a challenge as well as an opportunity to brand their cities such that they can reverberate with the need, demands and attractions of millennials. On this, while it has been argued that the youthful group is more oriented toward cultural experiences, the management should align their branding with digital solutions; especial digital infrastructures, as such are more compatible with this target group. Such a strategy, plus the integration of social media platforms would provide them a form of product differentiation and add some aura of uniqueness which serves the youths better. On the above, Ogg (2019) observed that over 87% of the millennials are inclined to trust social media for travel inspiration, and where such has been adopted, Ogg (2019) argues that it translates to approximately 46% of successful bookings and this justifies why social media should be part and parcel of the urban branding strategy. Starĉević and Konjikušić (2018) further justify the need for social media by noting that though millennials are price sensitive, they would be willing to strain their budget if the target destinations offer unique and memorable experiences that can be shared via social media platforms. Such prompt them to spend considerably longer times on the internet searching and comparing destinations that would satisfy their cultural experience desires at affordable pricing, irrespective of location.

While the economic contribution of the tourism sector in any economy is conventionally agreed, the role of technology therein, coupled with that of mixity can, however, lead to negative impacts. In particular, those of homogenizing the urban fabric if the branding are influenced by the popular modernist planning ideologies. This is possible as most data that are used in such branding exercises are sourced from different collected datasets that are then computed by Artificial Intelligence (AI) models; especially if such datasets emerge as a result of simplistic planning trends that are popular in this era. Such fears are affirmed by Li et al. (2016) who highlight that most modern cities are characterized of a built environment that resemble visually. According to them, these five main indicators are present: variance, cohesion, fragmentation, density and compactness.

Cities are observed to be turning towards homogeneity following trends such as demolition, destruction and rebuilding on old cultural architectural structures. This is a common phenomenon as usually those old architectural and urban artefacts are present in the middle of modern urban areas and current urban data supports high transient movement, leading urban leaders to succumb to modernist pressures. Such practices, though they yield desired outcomes from select groups, threatens the diversity and uniqueness and cultural aspects of most cities, thus, negatively impacting on their attractiveness—in particular to cultural tourism. This is supported by Meyer-Bisch (2013) who underscores that the homogenization of the city threatens the unique cultural identity and value and thus washes down the authenticity and integrity of those urban areas; characteristics that have been found to reverberate with cultural travelers in their quest for travel destinations. Following this, the New Urban Agenda, a product of UN Habitat III (2016), demands that urban developments are to piously respect, uphold and promote the culture and traditions of communities in cities. Such could be achieved by employing the power of digital technologies which have the capacity to help in upgrading cultural structures while at the same time, maintaining the desired uniqueness and the heritage associated with those structures.

## THE RISE OF CULTURAL TOURISM AND CONSUMPTION

During the past few decades, we have witnessed a myriad of disruptive technological advancements targeting global sectors like transport and communication, health, art and culture, tourism, and education to name a few. In particular, due to the increasing number of travelers warranted by the increase and popularity of both airlines and cruise ships, the tourism sector has greatly been transformed. For instance, in 2017, the International Airline Transport Authority (IATA 2018) report that the number of air travelers reached a recorded high of 4.1 billion people, marking an approximate 7.3% increase in the number of travelers recorded in 2016 by the same agency. Due to the increase in activities in this sector, the Global Air Traffic (Mazareanu 2019) projected that by the end of 2019, the number of travelers would increase by 500 million people to reach 4.6 billion. On water, the Cruise Line International Association (CLIA) (Kennedy 2019) recorded a significant increase in the use of cruise ship industry due to their attractive packages and activities, which builds towards an attractive cruise experience through the offer of entertainment options while being disconnected to the busy urban life. For example, in 2015, their users reached 23 million, but increased to approximately 26.7 million people by 2017 and these trends are expected to continue with travelers opting for this mode expected to beat the 32 million mark by end of year 2020. The emergence of cruise ships is seen to be particularly popular with millennials, as these are out to find new experiences which cruise ships make possible as they are seen to access remote areas more easily as compared to air travel. Here, despite the preferred mode of transport, the most significant conclusion is that those increasing numbers of travelers have the potential to translate to huge market for cities and urban areas that position themselves to offer the services and products that the travelers are yearning for.

The above reality is seen to be taken seriously in some quarters and as Postma et al. (2017) posit that cities are now responding to those numbers by creating new infrastructures and facilities that augur well with the demand of the traveler; be it in terms of communication, interaction, culinary experience, comfort and entertainment. For instance, it is not unusual to find diverse customized cuisines and restaurants located in same city targeting travelers searching for experience in particular cultures.

By doing this, such strategies ensure that the tourists maximize their stay in the cities as their experiences and demands are satisfied. Zukin (2009) highlighted that with proper infrastructures, especially the digital ones, cities are now able to offer travelers opportunities to do online shopping and quick delivery of the of their orders. That way, travelers are not forced to travel to cities for particular products that were associated with particular cities. On this, technology have allowed for products from different cultural backgrounds and geographical locations to be stocked in different cities or different shops at the same and at the same price and quality. What matters then, in such circumstances is for the city management to forge unique strategies that will ensure they attract travelers, and subsequently, retain them as much as possible.

However, in the quest of maximizing the economic potential associated with tourism and the travel industry, it is worth being wary of the negative impacts that may arise from the same. In particular, such are understood to have their toll on cultural heritage and on the innate identity of urban areas; more so due to the aspect of causing homogenous landscapes as discussed in the previous section. With such negatives, the value and pride of the cities, as expressed by different urban thinkers like Jane Jacobs (Jacobs 1961), Nikos Salingaros (Alexander et al. 1977) and Christopher Alexander (Seamon 2007; Alexander 2002) and others are seen to deteriorate, and in the long run, they compromise on the liveability, inclusivity and sustainability aspects of the city. With the uniqueness of the city gone, Hocaoğlu (2017) expresses that it becomes a daunting task for city managers to create a brand or image that will be identified with the city and promote its tourist attractiveness, and this, is not healthy for competitiveness.

Therefore, instead of taking the route of creating infrastructures that compromise the culture and uniqueness of the city, it is noted that integrating the new technologies in the already existing urban fabric can help preserve the uniqueness of the city, while at the same time improve the experience of upcoming travelers. This notion is supported by Genç (2017) who highlights that the use of technology provides a platform for city managers to showcase their cultural heritage through different forms enabled by technology. For instance, it is now possible to showcase cultural products, arts and other heritage in virtual forms, video format and graphically, which are more appealing to the millennials who form a substantial part of the modern travelers. By using technologies such as Virtual Reality (VR), Augmented Reality (AR) and others, most of the cultural aspects of cities can be preserved and enjoyed by different demographic groups, and this would not only help in increasing the revenue in the city, but would also go a long way in ensuring posterity of threatened cultural heritage. Jung et al. (2015) add that by using those new technologies, there is unlimited potential for cities to improve the value of their services, and allow for customization of the same such that travelers have the latitude to enjoy the cultural products in the form they like and at their own pleasure; thus, make their experience even better, and this can even prompt them to spend more to continue their cultural consumption.

## Urban Diversity and Technology

As global economic sectors gain traction warranted by technological advancement of technology, the travel industry has not been left behind and is seen to be experiencing positive growth, especially due to the growth in air travel. On this, it is now evident that air travel is playing a predominant role in linking different cities across the globe, and this influencing more people to travel to various and more remote destinations. Unlike before, air travel has enabled movement of more people who seek different satisfactions in business, education, adventure, tourism, entertainment, holidays and leisure to name a few. Additionally, it has brought notable impacts in sectors like tourism and hospitality industries where there is now more competition as stakeholders try to capture the enlarging market necessitated by the increasing number of travelers. In particular, as noted by Richards (2018), extra attention is now focused on millennials who are seen to be more active in travelling, especially in their quest to fulfil their desires for unique cultural experiences that different cities around the world possess. Having identified that cultural tourism is becoming popular among travelers, travel agencies, city managers and other stakeholders are seen to increase their effort by innovating and bringing on board refined, and customized cultural products for increased marketability. However, while such strategies are seen to yield unquestionable results, some of those are seen to negatively impact traditional and native urban heritage, more so through demolition and urban regeneration programs that target to align urban structures with modernist perspectives. Such impacts have the potential to influence creation of homogenous urban landscape that is counterproductive in respect to achieving authenticity, competitive advantage, uniqueness and innovation, as explained by Appendino (2017).

While those challenges are eminent when stringent measures are overlooked, it is now evident that such could also be overcome. This is evident on how cities have been observed to use available, modern technologies as tools for promoting their diverse cultural products; thus, attracting more tourists and promoting economic growth. A classic example of how technology has been used is showcased by the Guggenheim Museum that is located in Bilbao, and which underwent notable structural transformation, and in turn, its emergence influenced greatly the revitalization of the Bilbao city, which is reported to have been deteriorating. After the Museum was built, Plaza and Haarich (2013) express that it caused an astronomic growth of both local and foreign visitors in the city, and those have managed to help turn around the then stagnating economy of the city to make it vibrant and competitive. In Petersburg Kentucky, the Ark and Creation Museum is said to have brought significant changes in making the city an attractive destination for travelers. The museum is based on the Christian belief of creation, and through technology, such beliefs have been emphasized and given a whole new dimension that the visitors are able to identify with, especially in respect to animation of some aspects of creation characters (Ham 2017).

The above two examples demonstrate how cities are benefiting by integrating technology in highlighting their cultural strength, and such has brought tangible impact in the job market and influencing the quality aspects of the urban fabric. Such benefits, as Pietro et al. (2018) highlight are increase even more especially with the integration of disruptive mobile apps that have taken the branding concepts of cities a notch higher. With apps cities are now able to efficiently showcase and share crucial information like locations of their cultural products, their cultural and cultural events' calendars and background information of some of their cultural heritage. Also, some apps have the capacity to showcase live images of different parts of the cities, the maps, the distances between cultural sites and a myriad of other crucial information that cultural tourists may be interested in during their research for places to visit. Better still, interested visitors can do online bookings, reserve seats in theaters, ride sharing, schedule their programs among many other things in the comfort of their homes or hotel rooms (McKinsey & Company 2018). The good thing about targeting the use of mobile apps as cities' branding tool is that mobile devices, especially smartphones are now ubiquitous globally, and these have the capacity to connect via different internet source; be it from mobile service providers or from numerous Wi-Fi hotspots that are now available in cities. Thereby, mobile users are able to access information from apps in real-time.

On that breath, Jamaluddin (2017) expresses that they are numerous apps that are available targeting the promotion of cities which include *Eventbrite*; a customizable app that allow event enthusiasts to search for upcoming events in their areas. Through this app, it is also possible for promoters to sell virtual tickets, making the whole experience of promoting, searching and paying for events seamless, less time consuming and very comprehensive for both the promoters and those after the events. Another popular app that has brought value to the idea of searching for events is the *Field Trip* that is associated with Google (Ingraham 2012). This app integrates the power of Augmented Reality (AR) such that users not only get information on different cultural interest like food, architectural structures, and background histories among others displayed virtually, but the same is narrated to them via the headphones of their devices. *Vamos* is another powerful app that has disrupted the branding concepts with its ability to capture even events being promoted by other apps like Facebook, Eventbrite, Tricketmaster, *Gravy*, *SongKick Concerts*, *MapMyNearest*, and others ensuring that users are well versed with what is happening in their area of interest. These are just a few in the mobile app market and many more are expected to continue coming up as the concept of city branding continue gaining traction.

Besides using mobile apps, urban managers and stakeholders therein have also been active in employing new innovative ideas to ensure that their urban brand remains competitive. Following that, the world is amass with famous city brands like *I Love New York*, *State of COLORADO*, *"Chief Storyteller"* in Detroit, #IAMSTERDAM [37], the Luxembourg tag line *"Let's Make it Happen"*, *"The Wild Within"* for British Columbia

(British Columbia 2019) and "*My Neighborhood*" in Buenos Aires amongst unlimited others that cities are pursuing (The Place Brand Observer 2018). To go further, most cities, integrate their branding campaigns into social media platforms and through the use of captivating hashtags, they share their brand logs, interesting photos and images of their products and engage in discussions geared toward telling the superior quality of life of their cities. One commonality with these captivating branding strategies that cities have resulted to is that they contributed to successful stories that allow visitors to conceptualize the cities in various dimensions according to their liking. Another inference from the branding strategy is that each city is able to invoke a sense of desire for the travelers to travel in them, which ultimately would lead to the city benefiting from such visits.

While branding strategies may be seen to only focus on exposing the attractiveness and uniqueness of the target city, they are also meant to imprint on the cultural diversity of cities which residents and even visitors can associate with (Kavaratzis and Hatch 2013). On this, Cotîrlea (2014) highlights that through the said promotion, the cultural heritage of a city, even those that may be minute is able to emerge, and this recognition goes a long way in promoting social inclusivity in the city as everyone feels appreciated and valorized. By ensuring that everyone appreciates the cultural diversity in the city, and that none of the heritage is looked down on, cities become complexly tied together and, in the process, such qualities like resilience, liveability, attractiveness, innovation and sustainability are enhanced. However, Burgess (2007) points that as cities pursue to promote the multiplicity of cultural diversity, such should not be approached in a unilateral way, since, eventually, such would led to emergence of a homogenous culture where certain aspects of culture like language, religious beliefs and arts are replaced by homologous pursuits. Such are said to happen when new urban cultures are born due to the free crossbreeding of different groups of people who are not able to identify with any specific culture and have to craft their own cultural path. Hervé (2017) links this new approach to governance, interaction and social networking, entertainment and income generation methods and such, though feasible for that group of people have negative impacts on existing cultures. Therefore, while new urban cultures will always crop out, it is novel to capitalize on the power of new technologies to secure aspects of existing culture for posterity (Boboc et al. 2019).

## Autonomous and Homogenous Cities

Cities have been known to play pivotal roles in economic growth, promote social inclusivity and serve as epicenter for political powers among other things in any given national economy. And, in the recent past, in line with those roles, cities across the globe are on the move toward adopting the most novel and most advanced strategy; allowing them to maintain the momentum in competitiveness, through adoption of modern urban technologies. Such strategies are in part informed by the increase in urban challenges like urbanization and ever-increasing urban population, plus the looming impacts of climate change. Tzafestas (2018) and Cheng et al. (2018) among others acknowledge that cities are now better placed since they have access to a wide range of urban digital solution tools and technologies such as Big Data, IoT and AI that would help in coming up with informed management decisions, with minimal human interventions. With these technologies coupled with numerous smart devices and sensors, urban managers now have the opportunity to complement their decision making processes by relying on analysis of numerous urban data that these devices are able to collect, store and transmit to analytical platforms. Additionally, data is paramount in improving the quality of urban decision making. As noted by Alam et al. (2014), this can be sourced from urban residents and visitors, especially by relying on different social media platforms where most of such data are publicly shared. The availability of such data is enabled by the increasing use of mobile devices, which have become ubiquitous in modern days, and the capacity of such devices to connect to the internet allows them to share massive data into the networks. From the data shared from those devices, it is possible to gather information from almost every sector of the urban fabric and that is a plus noting that data has become a proponent in decision making processes. As shown in the above section, having substantial data, especially from visitors and residents alike would prove significant when deciding on the branding strategy to adopt, especially where such entails showcasing the cultural diversity that the city is endowed with.

From a wide range of emerging literature, it is being argued that adoption of the said technologies in urban areas has the potential to lead to creation of autonomous cities, where human decisions are greatly suppressed (Van Winden and van den Buuse 2017; Danigelis 2017). In particular, this is noted to be made possible by advancement in the AI and the increasing number of smart devices that have the capacity to integrate well with AI technologies. The notion of having autonomous cities is already invoking some excitement, especially due to its potential to render decision making real-time with little or no chances of bias as expressed by Kim et al. (2017). On this, Norman (2018), posits that with autonomous cities issues of sustainability positively impacted. Such cities also experience few hiccups in the transport sectors, in assuring security to both locals and visitors.

While the role of AI in bringing notable disruptions and transformation in urban areas have been hailed, there are some apprehensiveness on it being fully relied upon without any human input; at least for now. On this, there are loads of issues that have been raised and believed could go wrong if human interventions are completely erased. For instance, Dubey et al. (2016) insist that AI, has the potential, in some instances to overlook, neglect and in some cases, treat certain sectors by disregarding and overlooking other sectors which cannot be easily quantified. That is, it is possible for this technology to consider sectors like health, security, education and transport among others superior to such like culture and art, because the amount of data generated from these sectors differ. But, from the discussion in this chapter, it has been demonstrated that cultural aspects of a city, though having been neglected and overlooked for long have the potential to bring sustainable development in a city, and AI could overlook them due to limited data that may be generated from the sector.

Another issue raised against AI is its higher propensity to treat urban operations as homogenous, thus, lead to unilateral decisions. In such cases, the aspect of cultural diversity would be trampled and with such, any effort to regenerate cultural aspects in urban areas would hit a dead end. Therefore, even without going further, it is clear that human interventions, no matter how minute, are still required in the running of urban affairs, until a time when all issues of digitization and urban technologies are streamlined. For now, buying from the words of Zorins and Grabusts (2015) who holds that most technologies, more so those related to AI like Artificial Neural Networks (ANN)—which is seen as the most advanced as it is meant to mimic human neural system, is still in its infancy state, there is need to ensure human input are maintained in cities (Allam 2019c). The idea of human inputs also hold noting that technologies also have the potential to produce mechanical cities, and in such cases, such intricate facets like liveability, 'wholeness', and human relationship would not be promoted, but in their stead, disjointed and fragmented cosmetic beauty perpetuated by modernists would remain. On this front, while AI may prove superior in regard to speed, reliability, scalability, latency, volume and convergence in decision making as shown by He and Garcia (2009), it cannot compare to human capabilities when decisions and options on critical and contentious issues are in question. After all, it only relies on the dataset used to train it.

## References

Alam, J. R., Sajid, A., Talib, R., & Niaz, M. (2014). A Review of the Role of Big Data in Business. *International Journal of Computer Science and Mobile Computing, 3*, 446–453.

Alexander, C. (2002). *The Nature of Order: The Process of Creating Life*. Berkeley, CA: The Centre for Environmental Structure.

Alexander, C., Ishikawa, S., & Silverstein, M. (1977). *A Pattern Language*. New York: Oxford University Press.

Allam, Z. (2019c). The Emergence of Anti-Privacy and Control at the Nexus between the Concepts of Safe City and Smart City. *Smart Cities, 2*, 96–105.

Allam, Z. (2020a). *Cities and the Digital Revolution: Aligning Technology and Humanity*. Springer International Publishing.

Allam, Z. (2020b). Data as the New Driving Gears of Urbanization. In Z. Allam (Ed.), *Cities and the Digital Revolution: Aligning Technology and Humanity*. Cham: Springer International Publishing.

Allam, Z. (2020c). On Culture, Technology and Global Cities. In Z. Allam (Ed.), *Cities and the Digital Revolution: Aligning Technology and Humanity*. Cham: Springer International Publishing.

Appendino, F. (2017). Balancing Heritage Conservation and Sustainable Development—The Case of Bordeaux. *IOP, 245*, 1–11.

Barbier, J., Delaney, K., & France, N. (2017). Digital Cities: Building the New Public Infrastructure: Cities Show Proven Results through Digital Transformation. Cisco.

Boboc, R. G., Duguleană, M., Voinea, G.-D., Postelnicu, C.-C., Popovici, D.-M., & Carrozzino, M. (2019). Mobile Augmented Reality for Cultural Heritage: Following the Footsteps of Ovid among Different Locations in Europe. *Sustainability, 11*, 1–20.

Böcker, L., Van Amen, P., & Helbich, M. (2017). Elderly Travel Frequencies and Transport Mode Choices in Greater Rotterdam, the Netherlands. *Transportation, 44*, 831–852.

British Columbia. (2019). Ministry of Tourism, Arts and Culture Launches New Roadmap to Bolster Tourism as Economic Driver [Online]. *Destination British of Columbia*. Retrieved July 4, 2019, from https://www.destinationbc.ca/

news/ministry-of-tourism-arts-and-culture-launches-new-roadmap-to-bolster-tourism-as-economic-driver/.

Burgess, C. (2007). Multicultural Japan? Discourse and the 'Myth' of Homogeneity. *The Asia-Pacific Journal/Japan Focus, 5*, 1–25.

Cheng, S., Li, X., Zhihan, L. V., Song, H., Jia, T., & Lu, N. (2018). Data Analytics of Urban Fabric Metrics for Smart Cities. *Future Generation Computer Systems, 107*, C. https://dl.acm.org/toc/fgcs/2020/107/C

Cotîrlea, D. (2014). From Place Marketing to Place Branding within the Nation Branding Process: A Literature Review. *Ovidius University Annals, Economic Sciences Series, XIV, issue 2*, 297–302.

Danigelis, A. (2017). Smart City Market Growth Driven by Companies Like Schneider Electric and Ingersoll Rand [Online]. *Energy Manager Today*. Retrieved March 4, 2019, from https://www.energymanagertoday.com/smart-city-market-growth-driven-companies-like-schneider-electric-ingersoll-rand-0171629/.

Dubey, A., Naik, N., Parikh, D., Raskar, R., & Hidalgo, C. A. (2016). Deep Learning the City: Quantifying Urban Perception at a Global Scale.

Fromm, J. (2018). Transitioning Travel to The Millennial Market [Online]. *Forbes*. Retrieved July 2, 2019, from https://www.forbes.com/sites/jefffromm/2018/09/05/transitioning-travel-to-the-millennial-market/#5b6d6bfd3219.

Genç, R. (2017). The Impact of Augmented Reality (AR) Technology on Tourist Satisfaction. In T. Jung & M. C. T. Dieck (Eds.), *Augmentedd Reality and Virtual Reality, Progress in IS*. Springer International Publishing.

Ham, K. (2017). The Phenomenal Economic Impact of the Ark and Creation Museum [Online]. *Genesis Is the Answer*. Retrieved July 3, 2019, from https://answersingenesis.org/ministry-news/core-ministry/phenomenal-economic-impact-ark-and-creation-museum/.

He, H., & Garcia, E. A. (2009). Learning from Imbalanced Data. *IEEE Transactions on Knowledge and Data Engineering, 21*, 1260–1284.

Hervé, J. (2017). What's Trending? On the Cultural Challenges Facing Cities [Online]. *City Metric*. Julie Hervé. Retrieved July 3, 2019. https://www.city-metric.com/politics/what-s-trending-cultural-challenges-facing-cities-3103

Hocaoğlu, D. (2017). Challenges in Promoting Cities through Culture Within the New Global Economy. In Association, M (Ed.), *Advertising and Branding: Concepts, Methodologies, Tools, and Applications*. Hershey, PA: IGI Global.

IATA. (2018). Traveler Numbers Reach New Heights [Online]. *IATA*. Retrieved July 2, 2019, from https://www.iata.org/pressroom/pr/Pages/2018-09-06-01.aspx.

Ingraham, N. (2012). Google Releases 'Field Trip' App, a Location-Aware Guidebook to Your Surroundings [Online]. *The Verge*. Retrieved July 4, 2019, from https://www.theverge.com/2012/9/27/3418082/google-field-trip-location-based-guidebook.

Jacobs, J. (1961). *The Death and Life of Great American Cities*. New York, NY: Vintage Books.

Jamaluddin, A. (2017). 10 Mobile Apps to Find Upcoming Events & New Places of Interest [Online]. *HongKiat*. Retrieved July 3, 2019, from https://www.hongkiat.com/blog/apps-events-activities-around-you/.

Jung, T., Chung, N., & Leue, M. C. (2015). The Determinants of Recommendations to Use Augmented Reality Technologies: The Case of a Korea Theme Park. *Tourism Management, 49*, 75–86.

Kavaratzis, M., & Hatch, M. (2013). The Dynamics of Place Brands: An Identity-Based Approach to Place Branding Theory. *Marketing Theory, 13*(1), 69–86.

Kennedy, S. (2019). *2019 Cruise Trends & Industry Outlook*. Cruise Line International Association.

Kim, T.-H., Ramos, C., & Mohammed, S. (2017). Smart City and IoT. *Future Generation Computer Systems, 76*, 159–162.

Li, X., Lv, Z., Hijazi, I., Hongzan, J., Li, L., & Li, K. (2016). Assessment of Urban Fabric for Smart Cities. *IEEE Access, 4*, 373–382.

Mazareanu, E. (2019). Global Air Traffic—Scheduled Passengers 2004–2019 [Online]. *Statista*. Retrieved July 2, 2019, from https://www.statista.com/statistics/564717/airline-industry-passenger-traffic-globally/.

Mckinsey & Company. (2018). *Smart City Solutions: What Drives Citizen Adoption Around the Globe?* Mckinsey Center for Government.

Meyer-Bisch, P. (2013). Cultural Rights Within the Development Grammar. Agenda 21 for Culture. Barcelona: United Cities and Local Governments (UCLG).

Nakanishi, H., & Black, J. A. (2016). Travel Habit Creation of the Elderly and the Transition to Sustainable Transport: Exploratory Research Based on a Retrospective Survey. *International Journal of Sustainable Transportation, 10*, 604–616.

Norman, B. (2018). Are Autonomous Cities Our Urban Future? *Nature Communications, 9*, 2111.

Ogg, A. (2019). The Millennial: Travel Trends of the Largest Generation [Online]. *Vacation Access*. Retrieved July 2, 2019, from https://www.vaxvacationaccess.com/the-compass/The-Millennial-Travel-Trends-of-the-Largest-Generation/.

Pietro, L. D., Mugion, R. G., & Renzi, M. F. (2018). Heritage and Identity: Technology, Values and Visitors Experiences. *Journal of Heritage Tourism, 13*, 97–103.

Plaza, B., & Haarich, S. N. (2013). The Guggenheim Museum Bilbao: Between Regional Embeddedness and Global Networking. *European Planning Studies, 23*, 1456–1475.

Postma, A., Buda, D.-M., & Gugerell, K. (2017). The Future of City Tourism. *Journal of Tourism Futures, 3*, 95–101.

Seamon, D. (2007). Christopher Alexander and a Phenomenology of Wholeness. In *Christopher Alexander Annual Meeting of the Environmental Design Research Association*. Sacramento, CA: EDRA.

Starčević, S., & Konjikušić, S. (2018). Why Millenials as Digital Travelers Transformed Marketing Strategy in Tourism Industry. *TISC - Tourism International Scientific Conference Vrnjačka Banja, 3*(1), 221–240. Retrieved from http://www.tisc.rs/proceedings/index.php/hitmc/article/view/12

The Place Brand Observer. (2018). 31 Innovative Destination Marketing Campaigns and Place Branding Strategies to Serve as Inspiration in 2018 [Online]. *Place Brand Observer*. Retrieved July 3, 2019, from https://placebrandobserver.com/31-inspiring-destination-marketing-campaigns-place-branding-examples/.

Tzafestas, S. G. (2018). Synergy of IoT and AI in Modern Society: The Robotics and Automation Case. *Robotics & Automation Engineering Journal, 31*, 1–15.

UN Habitat III. (2016). Socio-Cultural Urban Frameworks. In: *Policy Paper*. Quito: UN Habitat.

Viale Pereira, G., Cunha, M. A., Lampoltshammer, T. J., Parycek, P., & Testa, M. G. (2017). Increasing Collaboration and Participation in Smart City Governance: A Cross-Case Analysis of Smart City Initiatives. *Information Technology for Development, 23*, 526–553.

World Tourism Organisation. (2012). *Global Report on Aviation—Responding to the Needs of New Tourism Markets and Destinations*. Madrid, Spain: UNWTO.

Zorins, A. & Grabusts, P. (2015). Artificial Neural Networks and Human Brain: Survey of Improvement Possibilities of Learning. In: *10th International Scientific and Practical Conference*, Rezekne, Latvia, pp. 228–231.

Zukin, S. (2009). Destination Culture: How Globalization Makes All Cities Look the Same. In: *Rethinking Cities and Communities: Urban Transition Before and During the Era of Globalisation*. Hartford, CT: Centre for Urban and Global Studies, Trinity College.